Blanco's Overview of Alpha-1 Antitrypsin Deficiency

Blanco's Overview of Alpha-1 Antitrypsin Deficiency

History, Biology, Pathophysiology, Related Diseases, Diagnosis and Treatment

Ignacio Blanco
Alpha-1 Antitrypsin Deficiency Spanish Registry (REDAAT), Spanish Society of Pneumology (SEPAR), Barcelona, Spain

ELSEVIER

ACADEMIC PRESS
An imprint of Elsevier

Academic Press is an imprint of Elsevier
125 London Wall, London EC2Y 5AS, United Kingdom
525 B Street, Suite 1800, San Diego, CA 92101-4495, United States
50 Hampshire Street, 5th Floor, Cambridge, MA 02139, United States
The Boulevard, Langford Lane, Kidlington, Oxford OX5 1GB, United Kingdom

Notices

Knowledge and best practice in this field are constantly changing. As new research and experience broaden our understanding, changes in research methods, professional practices, or medical treatment may become necessary.

Practitioners and researchers must always rely on their own experience and knowledge in evaluating and using any information, methods, compounds, or experiments described herein. In using such information or methods they should be mindful of their own safety and the safety of others, including parties for whom they have a professional responsibility.

To the fullest extent of the law, neither the Publisher nor the authors, contributors, or editors, assume any liability for any injury and/or damage to persons or property as a matter of products liability, negligence or otherwise, or from any use or operation of any methods, products, instructions, or ideas contained in the material herein.

Library of Congress Cataloging-in-Publication Data
A catalog record for this book is available from the Library of Congress

British Library Cataloguing-in-Publication Data
A catalogue record for this book is available from the British Library

ISBN: 978-0-12-809530-0

For information on all Academic Press publications visit our website at https://www.elsevier.com/books-and-journals

Publisher: Mica Haley
Acquisition Editor: Stacy Masucci
Editorial Project Manager: Sam Young
Production Project Manager: Chris Wortley
Designer: Matthew Limbert

Typeset by Thomson Digital

To my beloved wife Cristina, my daughters Inés and Jimena, my son Ignacio and his wife Irene, and my grandson Nacho, for their support and kindness.

To the memory of my always remembered parents, Ignacio and Blanca, and my beloved brother, Chema.

To the memory of the outstanding scientist, exceptional partner, and good friend of mine, Frederick Joseph de Serres, and his kind and intelligent wife, Christine Marie Covone.

To the memory of the ever admired leader, advocate and friend of all patients and researchers in the "Alpha-1" field, Mrs John W. Walsh.

To my sister Blanca, to Marina García, and to the Núñez-Blanco and Ramírez-Fueyo families, and to the friends who have always been by my side in good and bad times.

All of them, in different ways, provided me the strength and enthusiasm and the appropriate conditions to undertake the laborious task of writing this book.

Contents

Preface

There is abundant scientific information available on original articles, reviews, and monographs focused on specific aspects of the alpha-1 antitrypsin, but it is lacking an easy-to-read updated book integrating the various facets of this glycoprotein in health and disease.

Paraphrasing Albert Einstein (1879–1955), since "everything must be made as simple as possible, but not simpler," this manual tries to be a summary compendium of the current knowledge about the various aspects of the alpha-1 antitrypsin with the aim of providing a first aid to the medical student or doctor facing (as clinical, laboratory analyst, or basic researcher) the deficiency of this essential protein.

The book has been entitled "Blanco's Overview of Alpha-1 Antitrypsin Deficiency …" This title, apparently conceited, was proposed by the Senior Acquisitions Editor because the manual has been written by a single expert author to give uniformity to the volume, avoiding redundancy (or even contradictions) of concepts, tables, and figures. Therefore, the Author is at once meritorious and responsible for both the strengths and weaknesses of what he has written.

According to a Chinese saying: "There are two ways of spreading light, to be the lamp that emits, or the mirror that reflects it" (Lin Yutang, 1895–1976), it could be said that the light emitted by this book (hopefully, intense and bright!) comes partly from a modest lamp achieved by the Author after nearly 30 years of dedication to "Alphas," but mostly by the potent lamps of many other authors related to the Alpha-One-Foundation (AOF), the Alpha-1 antitrypsin International Registry of (AIR), and the Spanish Registry of Alpha-1 Antitrypsin Deficiency (REDAAT).

Anyway, with all my heart, I hope this book will finally contribute to improve the quality of life of those affected as well as the future generations of "Alphas," whose ultimate goal is achieving the "cure."

Acknowledgments

First, I would like to express my deep gratitude to all my "Alpha" patients. Many of them not only placed in me their confidence, but also gave me their blood and tissue samples for studies, and also taught me many valuable details about their disease.

My grateful thanks are extended to Dr. Carmen Rodriguez, Dr. Hortensia Canto, Dr. Eloy Fernandez, Professor Sabina Janciauskiene, and Professor Enrique Bustillo for their very valuable and varied contributions over the many years we worked together.

I thank especially the professional assistance provided by my niece Ana Ramirez (policy advisor for the Council of the European Union), and to the Spanish Society of Respiratory Diseases (SEPAR) for allowing and encouraging the dissemination of knowledge by means of this publication.

I also would like to thank the Elsevier Editors (Miss Stacy Masucci, Mr Chris Wortley and Mr. Samuel Young), and to the six anonymous referees who assessed this project, for having chosen me to perform this task, and for their positive comments, useful suggestions, and excellent technical support.

Finally, aware that I probably may have forgotten some names, I want to thank everyone who helped me to write this book.

Chapter 1

Alpha-1 Antitrypsin Deficiency: Introduction and History

1.1 INTRODUCTION

Human alpha-1 antitrypsin (AAT), also known as alpha-1 proteinase inhibitor (α1-Pi) and SERPINA*1* (serine protease inhibitor, group *A*, member *1*), is a circulating glycoprotein whose main function is to inhibit neutrophil elastase and other serine proteases in blood and tissues [1].

AAT is the most abundant circulating protease inhibitor existing in human body. This glycoprotein is mainly synthesized and secreted by hepatocytes, and its main physiological function is to protect the matrix proteins of the lungs (especially elastin) from the proteolytic effects of proteases (especially neutrophil elastase and proteinase 3) released by activated and dying neutrophils [1,2].

The AAT gene locus is located in chromosome 14, and it is activated by byproducts generated during inflammatory or infectious processes [3]. The gene has two alleles, which are transmitted from parents to children by recessive autosomal codominant Mendelian inheritance. Normal alleles, present in 85%–90% of individuals, are designated M, and therefore a normal individual shows an MM genotype. The most prevalent deficiency alleles are designated S and Z, and their prevalence in Caucasian populations ranges from 2% to 10% and 1% to 3%, respectively. Consequently, although about 50 "rare" and "null" alleles with very low prevalence have been described, the vast majority of genotypes result from combinations of M, S, and Z alleles, that is, MM (normal genotype present in about of 85%–95% people, expressing 100% of AAT), MS, SS, MZ, SZ, and ZZ (five deficiency genotypes present in virtually all of the remaining people, and expressing *grosso modo* 80, 60, 55, 40, and 15% of AAT, respectively) [4] (Fig. 1.1).

AAT deficiency is an autosomal codominant genetic condition that mainly affects Caucasians of European heritage (although occasionally it can also affect individuals of other races), with an estimated prevalence of the most common severe deficient genotype (i.e., Pi*ZZ) of 1:2000–5000 individuals in Northern, Western and Central Europe, and 1:5000–7000 subjects of European ancestry living in Canada, United States, Australia, and New Zealand [4].

Typically, this congenital condition is characterized by the production of a defective AAT that forms stable polymers, which are retained within hepatocytes,

Blanco's Overview of Alpha-1 Antitrypsin Deficiency

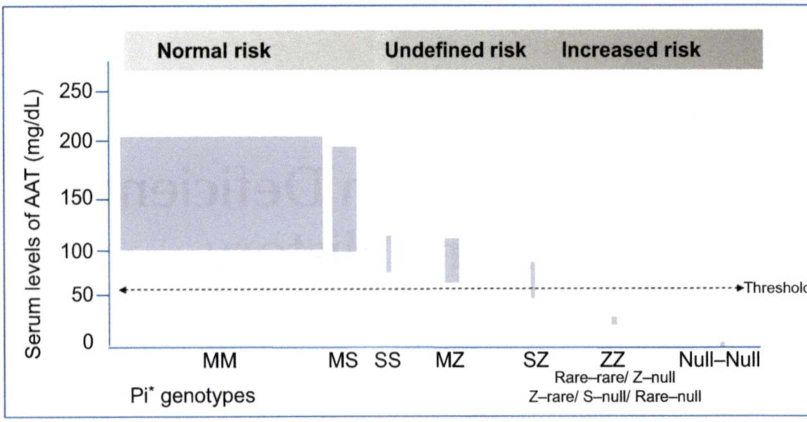

FIGURE 1.1 Alpha-1 antitrypsin (AAT) protease inhibitor (Pi*) genotypes, blood concentrations, and related risk for diseases development. The AAT protection level conventionally accepted is 57–60 mg/dL (or 11 μM). Below this cutoff *(dashed horizontal line)* are about of 10% SZ, all ZZ, and some combinations of Z and S with "null" and "rare" deficiency alleles. The size and thickness of the bars are roughly proportional to the prevalence of each genotype in Caucasians populations. In clinical practice, 96% of diseases are associated with ZZ genotype, and the remaining 4% with combinations of S, Z, "null," and severe deficiency "rare" alleles. *(Adapted from Blanco I, Fernández-Bustillo E, de Serres FJ, Alkassam D, Rodríguez MC. PI* S and PI* Z alpha 1-antitrypsin deficiency: estimated prevalence and number of deficient subjects in Spain. Med Clin. 2004;123:761–5.)*

secondarily leading to a decreased concentration and activity of AAT in blood and tissues. Severe AAT deficiency predisposes to early-onset of chronic obstructive pulmonary disease (COPD) in adults (typically pulmonary emphysema in cigarette smokers) and liver disease (predominantly chronic hepatitis and liver cirrhosis) in children and adults. Much less frequently, AAT deficiency is associated with systemic vasculitis, neutrophilic panniculitis and (not fully probed) to other inflammatory, autoimmune, and neoplastic diseases. Inexplicably only a part of the deficiency gene carriers manifests the disease, while between a third and a half are free of disease or manifest only minor symptoms throughout life [1–4].

While there is no specific treatment for AAT deficiency–induced liver cirrhosis, augmentation therapy with intravenous infusions of purified plasma-derived AAT obtained from serum of blood donors is currently available in some developed countries for lung-affected AAT-deficiency individuals [5].

1.2 HISTORY OF THE ALPHA-1 ANTITRYPSIN DEFICIENCY

The inhibitory capacity of proteases by human plasma was discovered by several authors in the late 19th century [6], but due to the insufficient development of laboratory techniques at that time, it was not possible to identify the responsible agent for this activity until 1955, when Herman Schultze (Germany) isolated the protein responsible of the antiprotease activity of the blood, which he called "α1-antitrypsin" for its location in the α1 band of globulins and its ability to inhibit pancreatic trypsin [7].

The fascinating history of the AAT deficiency began in 1952, when Carl-Bertil Laurell (1919–2001), head of the Laboratory of Clinical Chemistry at the University Hospital of Malmo (Sweden), introduced the electrophoresis of plasma proteins in his lab, and in 1962 (at an age of 43 years) he realized with this technique the absence of the α1 band in the protein electrophoresis of five samples. Although initially he thought it could be an artefact caused by bacterial contamination, he also found it suspicious that this accident was limited to the α1 band without affecting the rest of the bands. Thus, he requested to his rotary medical resident (then aged 30) Sten Eriksson (1932–) to seek clinical information of these patients, and Erickson found that three of them suffered from lung emphysema and had a loaded history of respiratory diseases, and even one of them belonged to a family with other family members affected by COPD. The two remaining patients had no AAT deficiency–related diseases, as often encountered in clinical practice with this condition [8] (Fig. 1.2).

The following year (1964), Eriksson detected two brothers with severe electrophoretic deficiency of α1 and lung emphysema, whose children had partial electrophoretic deficiency of α1 [9]. And (paraphrasing Louis Pasteur, 1822–95, as "in the field of observation, chance favors only the prepared mind") this finding led him to review the clinical histories from several Swedish hospitals, managing to collect 33 cases with AAT deficiency, the majority (n = 23) with COPD and several of these related each other. Then Eriksson made his doctoral thesis in 1965 on this subject (which he titled *Studies in Alpha-1 Antitrypsin Deficiency*), and together with his mentor Laurell, he concluded that AAT deficiency was an inherited disorder that favored the development of emphysema in early periods of adulthood [10,11].

In 1964, Gross et al. showed that intratracheal administration of papain (a cysteine protease present in the tropical plant papaya) induced lung emphysema in rats, suggesting a proteolytic mechanism in producing this pathology [12].

In 1967, Magne Fagerhol (Norway) noted that AAT had a polymorphic electrophoretic spectrum of presentation with several different electrophoretic patterns, and in accordance with Laurell, he proposed to call the set of AAT electrophoretic variants by the abbreviation Pi, which stands for *protease inhibitor*, system [13]. Initially, they designated the AAT variants with the alphabet letters, according to their rate of electrophoretic migration: Pi M (M of "medium") for those with medium speed, Pi S (of "slow") for those with slow migration rate, Pi F (of "fast") for those with the rapid migration, and Pi Z, the last letter of the alphabet, for those with very slow migration [14].

The association of AAT deficiency with liver cirrhosis was documented first in 10 children by Sharp (Minnesota, USA) in 1969, and he was the same author who also detected periodic acid Schiff–positive and diastase-resistant inclusions of AAT in cirrhotic livers for the first time [15].

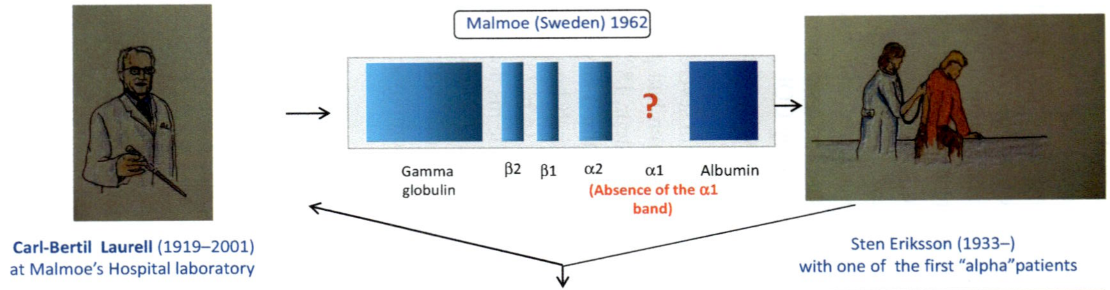

Clinical data from the five first α1- antitrypsin deficiency patients			
Cases	**Age/sex**	**Medical history**	**Diagnosis**
1	35/Female	COPD, pneumonia, neumothorax	Emphysema, bronquiectasis
2	38/Female	Pneumonía, COPD, two siblings with asthma and emphysema	Emphysema
3	44/Male	COPD, pneumonía	Emphysema, bronchiectasis
4	66/Female	Rheumatoid arthritis, osteoporosis	No respiratory disease
5	31/Female	Sinusitis	No respiratory disease

FIGURE 1.2 **The alpha-1 antitrypsin deficiency discovery: from an incidental laboratory finding to the fist "alpha-1 deficiency" patients.**

That same year, Turino and Snider (USA) related the development of pulmonary emphysema with neutrophil elastase [16], a serine protease discovered by Janoff and Scherer 2 years before [17], which is carried within the azurophilic (or primary) granules of neutrophils from the blood to the lungs (or other body tissues) where it can exert a potent elastolytic activity.

In 1972, Aaron Janoff also reported that AAT easily inhibited neutrophil elastase, and it was able to block the development of emphysema in animal models [18].

This same year, Berg and Eriksson described, for the first time, the association of AAT deficiency with adult liver cirrhosis [19].

In addition, in 1971–72 three French dermatologists: Warter, Grosshans, and Storck (Strasbourg) reported the first case of *family panniculitis with AAT deficiency* [20,21].

Over time, the laboratory diagnostic techniques were gradually improved, and since 1974 isoelectric focusing (IEF) became the standard technique for classifying AAT phenotypes [22]. (Currently, IEF can identify up to 30 AAT variants among the more than 100 described so far.)

In 1976, Owen and Carrell in Cambridge (United Kingdom) showed that the S variant was produced by the substitution of a molecule of glutamic acid for another of valine at position 264 of the amino acid chain [23], and 2 years later Jeppsson (Malmo, Sweden) reported that the Z variant was produced by the substitution of a glutamic acid by lysine at position 342 [24], this mutation causing serious conformational changes in the protein structure.

In 1976, Sveger published the results of a neonatal screening in 200,000 newborns from hospitals all over Sweden, detecting 127 Pi Z genotypes (mostly Pi ZZ, but a few Pi Z-null) [25]. The long-term medical follow-up from birth to present of this cohort has provided and continues to provide important data about the natural history of AAT deficiency [26–28].

In 1977–78 Larsson (Sweden) reported a close causal relationship between smoking habits and emphysema development [29,30].

In the 1980s, the structure of human AAT was revealed, its gene was cloned and sequenced, and its location was found to be situated at the distal end of the long arm of chromosome 14 [31–37].

In 1987, the US Food and Drug Administration (FDA), based on the demonstrated biochemical efficacy of AAT by the Gadek, Klein, Holland, and Crystal studies [38,39], approved the marketing of Prolastin (i.e., AAT obtained by Bayer Lab by fractionation and purification of healthy blood donors plasma) for augmentation therapy in adults with severe AAT deficiency–related emphysema [40–41]. Subsequently, the commercial firms of plasma derivatives: Baxter, Grifols, CSL Behring, and Kamada introduced their own products of purified AAT, obtained from pooled human serum, in the market of plasma derivatives, with the names of Aralast (2002), Zemaira (2003), Trypsone (2004), and Glassia (2010), respectively. More recently, CSL Behring has introduced its AAT in Europe under the name of Respreeza (2015) [42]. On the other hand, augmentation therapy has been available in France for almost 20 years, with the commercial name of Alfalastin [43].

In the 1990s, with the widespread use of DNA amplification techniques [i.e., polymerase chain reaction (PCR)] and sequencing of genes, the number of AAT variants has been increased to over 500, although only about 125 of these have been validated, including approximately 50 "rare" and "null" alleles among them [44,45].

Since the late 1990s, systemic vasculitis, especially those types showing circulating antineutrophil cytoplasmic antibodies, has been added to the list of diseases associated with AAT deficiency [1–3].

In late December 1991, David Lomas in Cambridge (United Kingdom) found that the Z protein has a marked tendency to spontaneously form stable polymers [46], and it soon became known that almost 90% of Z polymers are retained in the rough endoplasmic reticulum of hepatocytes, not being able to enter into the bloodstream, leading to low AAT concentrations in the plasma and tissues [47,48].

Following the discovery of AAT deficiency, and especially since the release of Prolastin for augmentation therapy, a progressively increased number of patient's registries and associations have been created in different countries, with the purpose of promoting diagnosis and research and to regulate the administration of augmentation therapy. The first one was the Danish Registry in 1978. After the Danish registry, many other registries and patient associations were created in several developed countries around the world. The American Registry was set up in the National Heart, Lung, and Blood Institute (NHLBI) in 1989 and stopped working in 1996 [49].

In the mid-1990s, John Walsh (USA), Sandy Lindsey and Susan Stanley, three people diagnosed with AATD, created the Alpha One Foundation (AOF) and AlphaNet, an exemplary medical–social system of health management involving affected people (popularly called "alphas") and their relatives. The AOF has collaborated with clinician-scientists to organize the Alpha One Foundation Research Network Registry, with the purpose of facilitating clinical research in AATD [50,51].

The Alpha One International Registry (AIR) was established in Malmo (Sweden) in 1997, with the aim of increasing knowledge about this condition, to make progress in basic and clinical research, and to implement multicenter clinical trials [52,53].

In 2003, a comprehensive guideline, based on the best available clinical evidence on diagnosis and treatment of AAT deficiency, was published by the American Thoracic Society (ATS) and the European Respiratory Society (ERS) [1], and other guides of similar content were developed in other several countries [54,55].

In latest years, there have been significant advances in genetic epidemiology [56–59], natural history, diagnosis, and detection of AAT deficiency, as well as in physiological and pathophysiological mechanisms [1–4]. Furthermore, the high-resolution computed tomography (HRCT), a highly sensitive method for measuring the density of lung, is being used alongside traditional methods of measurement of lung function to more reliably monitor the progress of emphysema in diverse randomized, double-blind, placebo-controlled trials designed to analyze the effectiveness of the augmentation therapy [60].

At the same time, in the last 10–15 years, several researchers have been identifying a growing number of inhibitory and noninhibitory properties of the AAT, which give the protein a broad number of relevant antiinflammatory, antimicrobial and immunomodulatory capacities. These facts are encouraging interest in the protein from clinicians, researchers, and pharmaceutical companies because of the potential possibilities for new clinical applications in

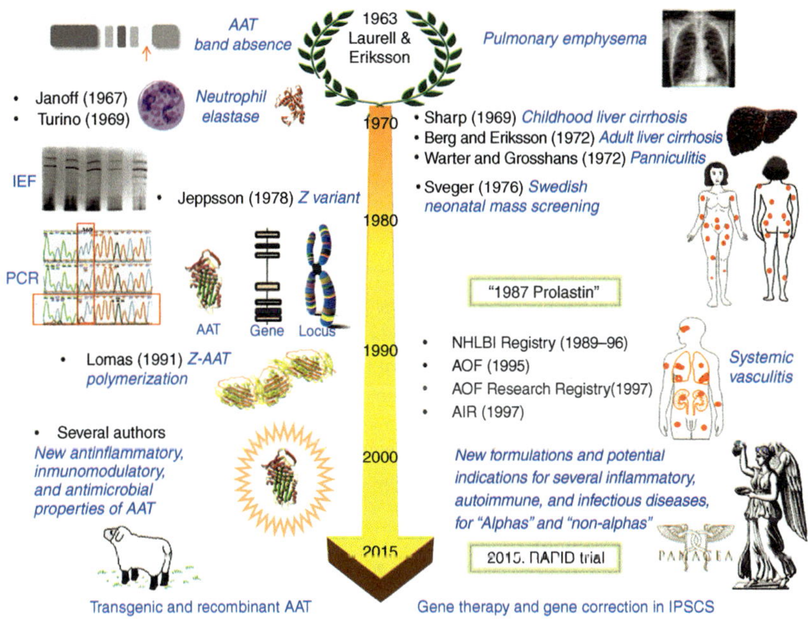

FIGURE 1.3 Graphic outline of the history of AAT deficiency. After the discovery of the of the AAT deficiency–related emphysema in 1963, it took 5–6 years to link emphysema with neutrophil elastase, and about a decade to discover its relationship with liver cirrhosis and panniculitis. In the next two decades, the laboratory diagnostic techniques were significantly improved. In 1987, the US Food and Drug Administration (FDA approved Prolastin for augmentation therapy. In 1992, Lomas discovered Z polymerization. Meanwhile, systemic vasculitis was added to the list of AAT deficiency–related diseases, and powerful foundations and patients registries were created with the aim of promoting research, diagnosis, and treatment. In latest years, researchers are identifying a growing number of AAT antiinflammatory and immunomodulatory properties with potential future applications for a number of diseases both inside and outside the AAT deficiency context (figuratively represented in the diagram by Panacea, "the goddess of universal health"). Currently, the attainment of recombinant AAT to be applied to humans seems already possible, and both gene and corrected pluripotent stem cell therapy have taken their first steps. *AIR*, Alpha One International Registry; *AOF*, Alpha One Foundation; *IEF*, isoelectric focusing; *IPSCS*, induced pluripotent stem cells; *NHLBI*, National Heart, Lung, and Blood Institute; *PCR*, polymerase chain reaction; *RAPID*, randomized, placebo-controlled trial in alpha-1 proteinase inhibitor deficiency.

inflammatory, autoimmune, and infectious diseases, both inside and outside the context of AAT deficiency [61–64].

In 2015, the results of the first study of high-methodological quality (i.e., RAPID trial) showed that augmentative therapy significantly reduces the loss of lung density in patients with emphysema associated with AAT deficiency [65], and other trials of similar design and objectives are currently underway.

On the other hand, the attainment of an adequate recombinant AAT to be applied to humans seems already close, and gene therapy and corrected pluripotent stem cell therapy have taken their first steps [66–71].

In summary, AAT deficiency is an exemplary model of solidarity, which has generated a loyal collaboration among patients and their families and clinical and basic scientists, nurses, therapists, public health professionals, and pharmaceutical companies. As a result of this cooperation, substantial progress has been made in some pivotal areas (Fig. 1.3). However, the AAT history is still incomplete and possibly will require a long way to go, with the ultimate goal of achieving the cure of patients and alleviating the concerns of their families in the near future. All these aspects mentioned earlier will be discussed in more detail throughout this book.

REFERENCES

[1] American Thoracic Society, European Respiratory Society. American Thoracic Society/European Respiratory Society statement: standards for the diagnosis and management of individuals with alpha1-antitrypsin deficiency. Am J Respir Crit Care Med 2003;168:818–900.

[2] Stoller JK, Aboussouan LS. Alpha1-antitrypsin deficiency. Lancet 2005;365:2225–36.

[3] Stoller JK, Aboussouan LS. A review of α1-antitrypsin deficiency. Am J Respir Crit Care Med 2012;185:246–59.

[4] de Serres F, Blanco I. Role of alpha-1 antitrypsin in human health and disease. J Intern Med 2014;276:311–35.

[5] Mohanka M, Khemasuwan D, Stoller JK. A review of augmentation therapy for alpha-1 antitrypsin deficiency. Expert Opin Biol Ther 2012;12:685–700.

[6] Fermi C, Pernosi L. Untersuchungen uber die enzyme. Veigerchende Studie Z Hig Infectionskr 1894;18:83–9.

[7] Schultze HE, Goilner I, Heide K, Schoenenberger M, Schwick G. Zur Kenntnis der alpha-globulin des menschlichen nirmal serums. Z Naturforsch 1995;10:463.

[8] Laurell CB, Eriksson S. The electrophoretic alpha1-globulin pattern of serum in alpha1-antitrypsin deficiency. Scand J Clin Lab Invest 1963;15:132–40.

[9] Eriksson S. Pulmonary emphysema and alpha-l-antitrypsin deficiency. Acta Med Scand 1964;175:197–205.

[10] Laurell CB, Eriksson S. The serum alpha-l-antitrypsin in families with hypo-alpha-l-antitrypsinemia. Clin Chim Acta 1965;11:395–8.

[11] Janciauskiene S. The discovery of alpha-1 antitrypsin. United Kingdom: Alpha-1 Awareness; 2010. Available from: http://www.alpha1awareness.org.uk/wp-content/uploads/2012/06/The-Discovery-of-AAT-AAW.pdf

[12] Gross P, Babyak ma, Tolker e E, Kaschak M. Enzymatically produced pulmonary emphysema; a preliminary report. J Occup Med 1964;6:481–4.

[13] Fagerhol MK, Laurell CB. The polymorphism of "prealbumins" and alpha-1-antitrypsin in human sera. Clin Chim Acta 1967;16:199–203.

[14] Fagerhol MK. The Pi-system, genetic variants of serum 1 antitrypsin. Series Haematol 1968;151:153–61.

[15] Sharp HL, Bridges RA, Krivit W, Freier EF. Cirrhosis associated with alpha-1-antitrypsin deficiency: a previously unrecognized inherited disorder. J Lab Clin Med 1969;73:934–9.

[16] Turino GM, Senior RM, Garg BD, Keller S, Levi MM, Mandl I. Serum elastase inhibitor deficiency and alpha 1-antitrypsin deficiency in patients with obstructive emphysema. Science 1969;165:709–11.

[17] Janoff A, Scherer J. Mediators of inflammation in leukocyte lysosomes. IX. Elastinolytic activity in granules of human polymorphonuclear leukocytes. J Exp Med 1968;128:1137–55.

[18] Janoff A. Inhibition of human granulocyte elastase by serum alpha-1 antitrypsin. Am Rev Respir Dis 1972;79:249–61.

[19] Berg NO, Eriksson S. Liver disease in adults with alpha-1-antitrypsin deficiency. N Engl J Med 1972;287:1264–7.

[20] Warter J, Storck D, Grosshans E, Metais P, Kuntz JL, Klumpp T. Syndrome de Weber-Christian associé à un déficit en alpha-I-antitrypsine. Enquéte familiale. Ann Med Interne 1972;123:877–82.

[21] Grosshans E, Warter J, Storck D. A propos d'une panniculite liquéfiante: les manifestations cutanées del déficits en alpha-I-antitrypsine. Bull Soc Fr Dermatol Syphiligr 1973;80:244–7.

[22] Allen RC, Harley RA, Talamo RC. A new method for determination of alpha-1-antitrypsin phenotypes using isoelectric focusing on polyacrylamide gel slabs. Am J Clin Pathol 1974;62: 732–9.

[23] Owen MC, Carrell RW, Brennan SO. The abnormality of the S variant of human alpha-1-antitrypsin. Biochim Biophys Acta 1976;453:257–61.

[24] Jeppsson JO, Laurell CB, Fagerhol M. Properties of isolated human alpha1-antitrypsins of Pi types M, S and Z. Eur J Biochem 1978;83:143–53.

[25] Sveger T. Liver disease in alpha1-antitrypsin deficiency detected by screening of 200,000 infants. N Engl J Med 1976;294:1316–21.

[26] Sveger T, Eriksson S. The liver in adolescents with alpha 1-antitrypsin deficiency. Hepatology 1995;22:514–7.

[27] Bernspång E, Wollmer P, Sveger T, Piitulainen E. Lung function in 30-year-old alpha-1-antitrypsin-deficient individuals. Respir Med 2009;103:861–5.

[28] Piitulainen E, Montero LC, Nystedt-Düzakin M, Stoel BC, Sveger T, Wollmer P, et al. Lung function and CT densitometry in subjects with alpha-1-antitrypsin deficiency and healthy controls at 35 years of age. COPD 2015;12:162–7.

[29] Larsson C, Eriksson S, Dirksen H. Smoking and intermediate alpha1-antitrypsin deficiency and lung function in middle-aged men. Br Med J 1977;2:922–5.

[30] Larsson C. Natural history and life expectancy in severe alpha1-antitrypsin deficiency, Pi Z. Acta Med Scand 1978;204:345–51.

[31] Kurachi K, Chandra T, Degen SJF, White TT, Marchioro TL, Woo SLC, et al. Cloning and sequence of cDNA coding for alpha-1-antitrypsin. Proc Natl Acad Sci USA 1981;78:6826–30.

[32] Lai EC, Kao F-T, Law ML, Woo SLC. Assignment of the alpha-1-antitrypsin gene and a sequence-related gene to human chromosome 14 by molecular hybridization. Am J Hum Genet 1983;35:385–92.

[33] Travis J, Salvesen GS. Human plasma proteinase inhibitors. Annu Rev Biochem 1983;52: 655–709.

[34] Carrell RW, Jeppsson JO, Laurell CB, Brennan SO, Owen MC, Vaughan L, et al. Structure and variation of human alpha 1-antitrypsin. Nature 1982;298:329–34.

[35] Long GL, Chandra T, Woo SLC, Davie EW, Kurachi K. Complete sequence of the cDNA for human alpha-1-antitrypsin and the gene for the S variant. Biochemistry 1984;23:4828–37.

[36] Darlington GJ, Astrin KH, Muirhead SP, Desnick RJ, Smith M. Assignment of human alpha-1-antitrypsin to chromosome 14 by somatic cell hybrid analysis. Proc Natl Acad Sci USA 1982;79:870–3.

[37] Perlino E, Cortese R, Ciliberto G. The human alpha 1-antitrypsin gene is transcribed from two different promoters in macrophages and hepatocytes. EMBO J 1987;6:2767–71.

[38] Gadek JE, Klein HG, Holland PV, Crystal RG. Replacement therapy of alpha 1-antitrypsin deficiency. Reversal of protease-antiprotease imbalance within the alveolar structures of PiZ subjects. J Clin Invest 1981;68:1158–65.

[39] Wewers MD, Casolaro MA, Sellers SE, Swayze SC, McPhaul KM, Wittes JT, et al. Replacement therapy for alpha 1-antitrypsin deficiency associated with emphysema. N Engl J Med 1987;316:1055–62.

[40] From the NIH: Intravenous replacement therapy for patients with severe alpha 1-antitrypsin deficiency. JAMA 1982;248:1683.

[41] Hubbard RC, Sellers S, Czerski D, Stephens L, Crystal RG. Biochemical efficacy and safety of monthly augmentation therapy for alpha 1-antitrypsin deficiency. JAMA 1988;260: 1259–64.

[42] Kueppers F. The role of augmentation therapy in alpha-1 antitrypsin deficiency. Curr Med Res Opin 2011;27:579–88.

[43] Gauvain C, Mornex JF, Pison C, Cuvelier A, Balduyck M, Pujazon MC, et al. Health-related quality of life in patients with alpha-1 antitrypsin deficiency: the French experience. COPD 2015;12(Suppl. 1):46–51.

[44] Bornhorst JA, Greene DN, Ashwood ER, Grenache DG. α1-Antitrypsin phenotypes and associated serum protein concentrations in a large clinical population. Chest 2013;143:1000–8.

[45] Lee JH, Brantly M. Molecular mechanisms of alpha1-antitrypsin null alleles. Respir Med 2000;94(Suppl. C):S7–S11.

[46] Lomas DA, Evans DL, Finch JT, Carrell RW. The mechanism of Z alpha 1-antitrypsin accumulation in the liver. Nature 1992;357:605–7.

[47] Carrell RW, Lomas DA. Alpha1-antitrypsin deficiency—a model for conformational diseases. N Engl J Med 2002;346:45–53.

[48] Ekeowa UI, Freeke J, Miranda E, Gooptu B, Bush MF, Pérez J, et al. Defining the mechanism of polymerization in the serpinopathies. Proc Natl Acad Sci USA 2010;107:17146–51.

[49] Survival and FEV1 decline in individuals with severe deficiency of alpha-1-antitrypsin. The Alpha-1-Antitrypsin Deficiency Registry Study Group. Am J Respir Crit Care Med 1998;158: 49–59.

[50] Walsh JW, Snider GL, Stoller JK. A review of the alpha-1 foundation: its formation, impact, and critical success factors. Respir Care 2006;51:526–31.

[51] Walsh JW. How patients learned to control their own future. COPD 2013;10:57–9.

[52] Stockley RA, Luisetti M, Miravitlles M, Piitulainen E, Fernandez P. Alpha One International Registry (AIR) group. Ongoing research in Europe: Alpha One International Registry (AIR) objectives and development. Eur Respir J 2007;29:582–6.

[53] Stockley RA, Dirksen A, Stolk J. Alpha-1 antitrypsin deficiency: the European experience. COPD 2013;10 Suppl. 1:50–3.

[54] Lara B, de la Roza C, Vilà S, Vidal R, Miravitlles M. Development and results of the Spanish registry of patients with alpha-1-antitrypsin deficiency. Int J Chron Obstruct Pulmon Dis 2007;2:393–8.

[55] Marciniuk DD, Hernandez P, Balter M, Bourbeau J, Chapman KR, Ford GT, Canadian Thoracic Society COPD Clinical Assembly Alpha-1 Antitrypsin Deficiency Expert Working Group. et al. Alpha-1 antitrypsin deficiency targeted testing and augmentation therapy: a Canadian Thoracic Society clinical practice guideline. Can Respir J 2012;19:109–16.

[56] Hutchison DC. Alpha 1-antitrypsin deficiency in Europe: geographical distribution of Pi types S and Z. Respir Med 1998;92:367–77.

[57] de Serres FJ. Worldwide racial and ethnic distribution of alpha1-antitrypsin deficiency: summary of an analysis of published genetic epidemiologic surveys. Chest 2002;122:1818–29.

[58] Blanco I, de Serres FJ, Cárcaba V, Lara B, Fernández-Bustillo E. Alpha-1 antitrypsin deficiency PI*Z and PI*S gene frequency distribution using on maps of the world by an inverse distance weighting (IDW) multivariate interpolation method. Hepat Mon 2012; 12(10 HCC):e7434.

[59] de Serres FJ, Blanco I. Prevalence of α1-antitrypsin deficiency alleles PI*S and PI*Z worldwide and effective screening for each of the five phenotypic classes PI*MS, PI*MZ, PI*SS, PI*SZ, and PI*ZZ: a comprehensive review. Ther Adv Respir Dis 2012;6:277–95.

[60] Stockley RA, Parr DG, Piitulainen E, Stolk J, Stoel BC, Dirksen A. Therapeutic efficacy of -1 antitrypsin augmentation therapy on the loss of lung tissue: an integrated analysis of 2 randomised clinical trials using computed tomography densitometry. Respir Res 2010;11: 136–44.

[61] Jonigk D, Al-Omari M, Maegel L, Müller M, Izykowski N, Hong J, et al. Anti-inflammatory and immunomodulatory properties of α1-antitrypsin without inhibition of elastase. Proc Natl Acad Sci USA 2013;110:15007–12.

[62] Lewis EC. Expanding the clinical indications for α(1)-antitrypsin therapy. Mol Med 2012;18:957 70.

[63] Blanco I, Lara B, de Serres F. Efficacy of alpha1-antitrypsin augmentation therapy in conditions other than pulmonary emphysema. Orphanet J Rare Dis 2011;6:14.

[64] Wanner A, Arce AD, Pardee E. Novel therapeutic uses of alpha-1 antitrypsin: a window to the future. COPD 2012;9:583–8.

[65] Chapman KR, Burdon JG, Piitulainen E, Sandhaus RA, Seersholm N, Stocks JM, et al. Intravenous augmentation treatment and lung density in severe α1 antitrypsin deficiency (RAPID): a randomised, double-blind, placebo-controlled trial. Lancet 2015;386:360–8.

[66] Ross D, Brown T, Harper R, Pamarthi M, Nixon J, Bromirski J, et al. Production and characterization of a novel human recombinant alpha-1-antitrypsin in PER.C6 cells. J Biotechnol 2012;162:262–73.

[67] Wang Z, Hilder TL, van der Drift K, Sloan J, Wee K. Structural characterization of recombinant alpha-1-antitrypsin expressed in a human cell line. Anal Biochem 2013;437:20–8.

[68] Kaushal S, Annamali M, Blomenkamp K, Rudnick D, Halloran D, Brunt EM, et al. Rapamycin reduces intrahepatic alpha-1-antitrypsin mutant Z protein polymers and liver injury in a mouse model. Exp Biol Med 2010;235:700–9.

[69] Hidvegi T, Ewing M, Hale P, Dippold C, Beckett C, Kemp C, et al. An autophagy-enhancing drug promotes degradation of mutant alpha1-antitrypsin Z and reduces hepatic fibrosis. Science 2010;329:229–32.

[70] Mueller C, Flotte TR. Gene-based therapy for alpha-1 antitrypsin deficiency. COPD 2013;10(Suppl. 1):44–9.

[71] Yusa K, Rashid ST, Strick-Marchand H, Varela I, Liu PQ, Paschon DE, et al. Targeted gene correction of α1-antitrypsin deficiency in induced pluripotent stem cells. Nature 2011;478: 391–4.

Chapter 2

Serpins and Serpinopathies

2.1 INTRODUCTION

Human alpha-1 antitrypsin (AAT), also called alpha-1 proteinase inhibitor and SERPINA1 (serine peptidase inhibitor, clade **A**, member **1**), is one of the most representative members of the serpins superfamily, a group of proteins with similar structures that were first identified by their ability to inhibit proteases, thereby maintaining the physiological protease–antiprotease balance to avoid the destructive proteolytic effects derived from an excess of free proteases.

In living beings there is a strict balance between proteases and antiproteases. Microorganisms and cells (naturally, and much more when they are activated by inflammation or cancer) release proteases to perform diverse physiological or pathological actions, and according to the composition of their active site (which determines the reactive substrate), proteases are, in simplified way, generally classified as serine proteases, cysteine proteases, aspartic proteases, and metalloproteases. To counteract them, the body has several classes of antiproteases. The most abundant are serpins, which, as their name indicates, have the capacity to neutralize serine proteases (or serine endopeptidases). The human tetrameric alpha2-macroglobulin acts as a panantiprotease able to inactivate, or to capture without inhibiting but preventing its action, a variety of proteases, including serine proteases, cysteine proteases, aspartic proteases, and metalloproteases. Finally, tissue inhibitors of metalloproteases specifically neutralize matrix metalloproteinases (Fig. 2.1). The detailed description of all these aspects exceeds the limits planned for this book, whose main objective is AAT deficiency.

2.2 THE SERPIN SUPERFAMILY

2.2.1 Classification and Nomenclature

*Ser*ine *p*roteinase *in*hibitors (serpins) are a very large and growing superfamily of proteins, consisting of approximately 1500 members sharing closely similar tertiary structures. The term serpin was proposed by Carrell and Travis in 1985 to describe the protein inhibitors superfamily of serine proteinases [1]. It is currently known that they are critical components of many physiological processes, including digestion, coagulation, fibrinolysis, complement regulation,

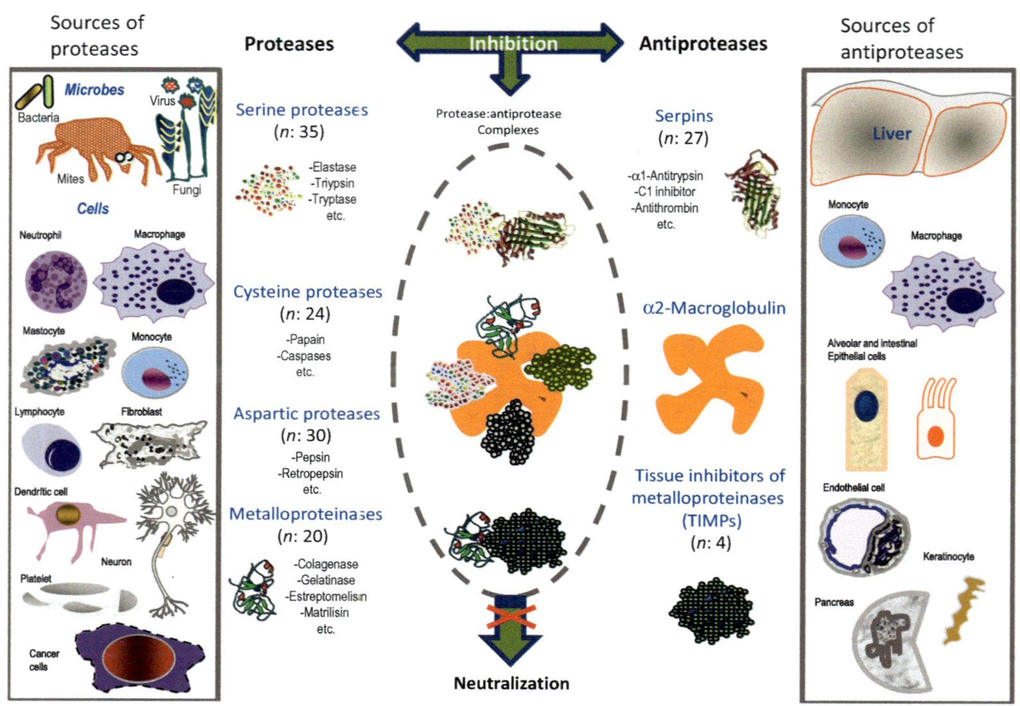

FIGURE 2.1 The protease/antiprotease balance. Simplified schematic representation of proteases and antiproteases (and their sources) involved in maintaining the physiological balance between them.

inflammation, apoptosis, angiogenesis, tumor suppression, hormone transport, and probably other [2].

Present in all living beings of the animal and plant kingdom (vertebrates, invertebrates, plants, and microbes) they have been classified into 16 different clades, lettered A–P, with numbers and letters assigned to them after the implementation of the Queen's College international consensus resolutions (Cambridge, 1999) to unify criteria for designation [3].

Human serpins, numbering 35, are in the first 9 clades, and lettered A–I (Table 2.1). Although 80% of human serpins perform fundamentally extracellular roles, and most of them are protease inhibitors playing a key role in controlling various proteolytic cascades essential for survival, other human serpins perform valuable no-inhibitory functions, such as proteins storage (e.g., albumin), hormone transport (e.g., thyroxine and cortisol), hormone precursors (e.g., angiotensinogen), or tumor gene expression suppression (e.g., maspin) (Table 2.2).

Human AAT is the most abundant circulating protease inhibitor and the prototypical representative of the serpin superfamily. For historical reasons, AAT

TABLE 2.1 Classification of Human Serine Protease Inhibitors (Serpins)

Clade	Letter	Names
Alpha-1 proteinase inhibitor, antitrypsin	A	SERPIN**A**1, SERPIN**A**2, SERPIN**A**3, SERPIN**A**4, SERPIN**A**5, SERPIN**A**6, SERPIN**A**7, SERPIN**A**8, SERPIN**A**9, SERPIN**A**10, SERPIN**A**11, SERPIN**A**12, SERPIN**A**13
Ovalbumin	B	SERPIN**B**1, SERPIN**B**2, SERPIN**B**3, SERPIN**B**4, SERPIN**B**5, SERPIN**B**6, SERPIN**B**7, SERPIN**B**8, SERPIN**B**9, SERPIN**B**10, SERPIN**B**11, SERPIN**B**12, SERPIN**B**13
Antithrombin	C	SERPIN**C**1
Heparin	D	SERPIN**D**1
Nexin, plasminogen activator inhibitor type I	E	SERPIN**E**1, SERPIN**E**2
Alpha-2 antiplasmin	F	SERPIN**F**1, SERPIN**F**2
C1 inhibitor	G	SERPIN**G**1
Heat shock protein 47	H	SERPIN**H**1, SERPIN**H**2
Neuroserpin	I	SERPIN**I**1, SERPIN**I**2

According to the Queen's College international consensus resolutions (Cambridge, 1999) serpins have been classified into 16 different clades, lettered A–P. Human serpins are in the first 9 clades, lettered A–I. The bold letter and the number indicate, respectively, the clade and gene number within the clade. Human alpha-1 antitrypsin (SERPIN**A**1) is the prototypical representative of the serpin superfamily. There are 17 other unclassified serpin groups (letters: J, K, L, M, N, O, P) found in animals, plants, insects, nematodes, and viruses.
Source: Modified from Silverman GA, Bird PI, Carrell RW, Church FC, Coughlin PB, Gettins PG, et al. The serpins are an expanding superfamily of structurally similar but functionally diverse proteins. Evolution, mechanism of inhibition, novel functions, and a revised nomenclature. J Biol Chem 2001;276:33293–6 [3].

TABLE 2.2 General Functions of Some Human Serpins

Serpin	Target enzyme	Main function
α1-PI (antitrypsin)	Neutrophil elastase, proteinase-3	Extracellular matrix remodeling
α1-Antichymotrypsin	Cathepsin G, chymase	Extracellular matrix remodeling, coagulation cascade, and inflammatory response modulation
α1-AP	Plasmin	Coagulation cascade (fibrinolysis)
PAI-1	u-PA, t-PA	Extracellular matrix remodeling, intracellular proteolysis, tumor suppression, and antimicrobial
PAI-2	u-PA, t-PA	Extracellular matrix remodeling and tumor suppression
AT III	Thrombin	Inhibits thrombin as well as other activated serine proteases regulating the blood coagulation cascade
PCI	Activated protein C	Limits the anticoagulant activity of protein C
C1-inhibitor	C1 esterase	Complement system inhibition
PN-1	Thrombin, uPA, plasmin	Extracellular matrix remodeling and prohormones conversion
Kalistatin	Kallikreins	Tissue kallikrein inhibitor, blood pressure regulation, and vascular remodeling

α1-PI, Alpha-1 protease inhibitor; AP, antiplasmin; AT III, antithrombin III; PAI, plasminogen activator inhibitor; PCI, protein C inhibitor; PN-1, protease nexin-1; u-PA, urokinase-type plasminogen activator; t-PA, tissue plasminogen activator.
Source: Modified from Potempa J, Korzus E, Travis J. The serpin superfamily of proteinase inhibitors: structure, function, and regulation. J Biol Chem 1994;269:15957–60 [5].

has been named SERPIN**A**1 (the bold letter **A** and the number 1 indicate the clade and gene number within the clade, respectively). Other members of the serpin superfamily include α1-antichymotrypsin (SERPIN**A**3), C1 inhibitor (SERPIN**G**1), antithrombin (SERPIN**C**1), and plasminogen activator inhibitor-1 (SERPIN**A**5), all of these playing an important biological role in the control of proteinases implicated in the inflammatory, complement, coagulation, and fibrinolytic pathways, respectively [4–6].

2.2.2 Structure and Inhibitory Mechanisms

Serpins are glycoproteins consisting of a central chain of 350–500 amino acids and several carbohydrate side chains, folding into 3 β sheets and 8 9 α helices, which hold a mobile loop of 20 residues that support the active site of the enzyme which serves as a substrate to the catalytic triad of serine proteinases [7].

Some serpins can take latent conformations, called **δ** (delta), with the active site fallen down and not exposed to proteinases. From this latent position, the serpin may expel the active site and become functional. Latency is common in antithrombin (SERPIN**C**1) and plasminogen activator inhibitor-1 (SERPIN**E**1), but only occasionally found in AAT (SERPIN**A**1), antichymotrypsin (SERPINA3), and neuroserpin (SERPIN**I**1) [8,9].

Briefly, the serpin active site acts as a decoy for the proteinase, which catches and cleaves it. This cleavage is followed by a dramatic conformational change in the serpin, which, as a tension state called "S" (from "stressed"), changes to one of relaxation called "R" (from "relaxed"). Then, the broken loop of the serpin abruptly inserts between the sheets of the trunk, and the captured protease is pulled abruptly to the opposite pole of the serpin, for a similar "trap to catch mice" or "catapult" mechanism. In the inhibitory pathway (e.g., applicable to the interaction between AAT and neutrophil elastase), both molecules become irreversibly inactivated [7,10]. However, in other cases, the proteinase can avoid the final phase of these events [3], and while the serpin results irreversibly inactivated, it remains free and active to react with other serpins (Fig. 2.2).

2.2.3 Polymerization

Some mutations of serpins with inhibitory activity generate unstable proteins with metastable conformations (M*), characterized by a global widening of the molecule and an abnormal fall of the active site. Molecules with these conformational changes tend to aggregate with other mutated molecules, forming dimers and polymers in the rough endoplasmic reticulum of synthesizer cells [10].

2.2.4 Serpinopathies

Several diseases caused by polymerization of some mutant serpins are grouped under the heading of serpinopathies, a generic term that includes a growing number of diseases whose common characteristic is the formation of polymers into the synthesizing cells of mutated serpins. The accumulation of polymers in the rough endoplasmic reticulum triggers cell stress and eventually irreversible cell damage, depending on the amount of protein accumulated (e.g., small amounts of polymers can be removed by the normal mechanisms of cell cleaning, and they are no harmful for the secreting cells). The retention of misfolded proteins within the synthesizing cells secondarily reduces its extracellular concentration, which may be insufficient to protect tissues from nonneutralized proteases. This latter effect is aggravated by the expression of serpins with marked conformational changes that significantly reduce their inhibitory capacity (Table 2.3).

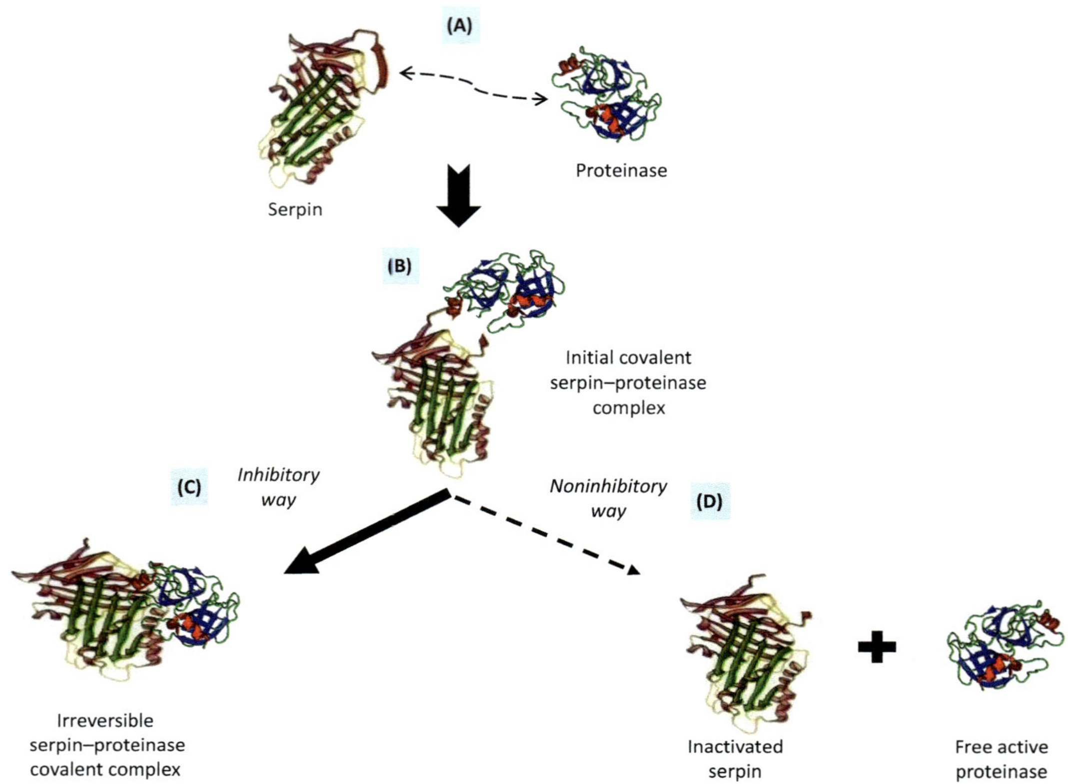

FIGURE 2.2 **Scheme of the serpin:protease inhibitory and no-inhibitory pathways.** The serpin active site acts as a decoy for the proteinase (A), which catches and cleaves it (B). In the inhibitory pathway, both molecules become irreversibly inactivated (C), as it occurs with the AAT and neutrophil elastase. In other cases, the proteinase can avoid the final phase of these events, and while the serpin results irreversibly inactivated, it remains free and active to react with other serpins (D).

TABLE 2.3 Serpinopathies, List of Diseases Related to Polymerization of Mutant Serpins

Serpin	Organs in which polymers accumulate (amount)*	Related diseases and mechanisms	Availability of augmentation therapy
α1-Antitrypsin (SERPINA1)	Liver (+++)	Liver cirrhosis by Z-AAT polymers accumulation; lung emphysema by excess of nonneutralized elastase	Yes
Antithrombin (SERPINC1)	Liver (+)	Vascular thrombosis by excess of factors IX and X and thrombin	Yes
C1-inhibitor (SERPING1)	Liver (+)	Angioedema by excess of C1r and C1s hydrolyzing C1 of the complement system, and nonneutralized kallikrein which releases bradykinin	Yes
α1-Antichymotrypsin (SERPINA3)	Liver (+)	Emphysema by excess of neutrophil cathepsin G and mast cell chymase; dubious relationship with Parkinson's and Alzheimer's by increased amyloid fibril deposits in CNS	No
Neuroserpin (SERPINI1)	Brain (+++)	Dementia [familial encephalopathy by neuroserpin inclusion bodies in neurons (FENIB)] and epilepsy by excess of free tissue plasminogen activator (t-PA)	No
Transcortin (SERPINA6)	Liver (+)	Hypocortisolemia. Possible relationship with hypo/hypertension and chronic fatigue syndrome (not well established)	No
Thyroxine-binding globulin (SERPINA7)	Liver (+)	Hypothyroidism due to reduced transport of thyroxine (T4) and triiodothyronine (T3) from blood to tissues	No

Mutant alleles of alpha-1 antitrypsin, antithrombin, C1-inhibitor, and alpha-1 antichymotrypsin favor the spontaneous formation of polymers and the retention of proteins within hepatocytes. But while liver AAT protein production is high and therefore the accumulated amount of polymers very large, a fact which favors the development of liver disease, the accumulated amount of antithrombin, C1-inhibitor, and alpha-1 antichymotrypsin is small and liver can easily clear their deposits avoiding cell damage. The lack of plasma AAT, antithrombin, and C1-inhibitor can result in emphysema, thrombosis, and angioedema, respectively. Polymerization of mutants of neuroserpin results in the retention of polymers within neurons causing familial encephalopathy with neuroserpin inclusion bodies (FENIB). Finally, polymerization of transcortin- and thyroxine-binding globulin serpins might theoretically cause deficiency of these hormones, but this has not been absolutely demonstrated so far.

*+, *Little*; +++, *large*.

2.2.4.1 Alpha-1 Antitrypsin Deficiency

AAT [SERPINA1, serpin peptidase inhibitor, clade A (alpha-1 antiproteinase, antitrypsin), member 1] deficiency is the main topic of this book, and will be more widely discussed in other chapters.

2.2.4.2 Alpha-1 Antichymotrypsin Deficiency

The gene for alpha-1 antichymotrypsin [SERPINA3, serpin peptidase inhibitor, clade A (alpha-1 antiproteinase, antitrypsin), member 3] shares locus with the AAT in the distal end of the long arm of chromosome 14, and is expressed primarily in liver and pancreas. Its function is to inhibit the neutrophil serine proteinase cathepsin G as well as mast cell chymase. Although more studies are needed to establish its relationship with clinical diseases, it has been suspected that SERPINA3 deficiency could favor the development of COPD, Parkinson's and Alzheimer's disease [11,12].

2.2.4.3 Antithrombin Deficiency

The gene for antithrombin [SERPINC1, serpin peptidase inhibitor, clade C (antithrombin), member 1] is expressed predominantly in liver. Antithrombin function consists of modulating the coagulation cascade inhibiting the excess of factors IX and X, as well as thrombin.

Antithrombin hereditary deficiency has a prevalence of 1:2000–1:5000 subjects, and can cause deep vein thromboses and pulmonary thromboembolisms. There are two types of antithrombin deficiency: type I and type II. Type I is characterized by a decrease of activity and concentration of this serpin, while type II is characterized by a decreased activity of this serpin (below 50%) but normal serum levels.

Antithrombin replacement therapy is indicated to prevent and treat thromboembolic complications associated with the deficiency of this serpin [13].

2.2.4.4 C1 Inhibitor Deficiency

The C1 inhibitor gene [SERPING1, serpin peptidase inhibitor, clade G (C1 inhibitor), member 1] is mainly expressed in liver, and with less intensity in adipocytes, lung, and ovary tissues. The protein works inhibiting the C1 proteinase complex of the complement system, preventing the abnormal activation of this cascade. In addition, it neutralizes serum kallikrein, preventing the formation of the inflammatory mediator bradykinin, a vasodilator peptide which dilates blood vessels via release of prostacyclin, nitric oxide, and endothelium-derived hyperpolarizing factor.

Congenital deficiency of C1 inhibitor is very rare, with an estimated prevalence of 1:10,000–1:50,000. It predisposes to the development of hereditary angioedema, autoimmune diseases (especially systemic lupus erythematosus), and age-related macular degeneration [14]. In clinical practice, 85% of cases show low serum levels of C1 inhibitor, and 15% normal but dysfunctional levels.

In some developed countries, it has been approved the use of replacement therapy with C1 inhibitor obtained from purified plasma of healthy donors to treat patients with angioedema caused by this rare congenital deficiency [15].

2.2.4.5 Neuroserpin Deficiency

Neuroserpin [SERPINI1, serpin peptidase inhibitor, clade I (neuroserpin), member 1] is a serpin secreted by neurons which participates in the axonal growth regulation of both central and peripheral nervous system neurons, as well as in synapses development by changing the number or strength of connections between brain cells for learning and memory. This serpin acts inhibiting urokinase, plasminogen, and plasmin activation, three different serine proteinases involved in fibrinolysis and extracellular matrix degradation.

Five types of familial mutations have been described so far; all characterized by neuroserpin polymers deposits within neurons causing familial encephalopathy with neuroserpin inclusion bodies. Familial encephalopathy with neuroserpin inclusion bodies express clinically by dementia, epilepsy, and trembling in different stages of life [7,16].

2.2.4.6 Transcortin Deficiency

Transcortin [SERPINA6, serpin peptidase inhibitor clade A, member 6, also corticosteroid-binding globulin] shares gene locus with AAT on chromosome 14. It is expressed in liver, and encodes a protein whose function is the transport of steroid hormones (especially cortisol) by blood. Transporting ligands is performed with the serpin in "S" conformation, but when it contacts tissues is cleaved by proteases adopting the "R" form and releasing its ligands locally. The highest concentration of proteases in the inflammatory focus makes much faster and intense the release of cortisol there [17].

Transcortin congenital deficiency is a very rare disorder, and its consequences little known. This disorder is characterized by low (or normal but dysfunctional) plasma transcortin and cortisol concentrations. Transcortin deficiency has been associated with abnormal changes in blood pressure and chronic fatigue syndrome, although conclusive studies are necessary to confirm or rule out whether it has clinical implications [18].

2.2.4.7 Thyroxine-Binding Globulin Deficiency

Thyroxine-binding globulin [SERPINA7, thyroxine-binding globulin, serpin peptidase inhibitor, clade A (alpha-1 antiproteinase, antitrypsin), member 7] is secreted by the liver, and together to prealbumin and albumin is one of the proteins carrying thyroid hormones from blood to tissues. Transport is performed with the serpin in conformation "S," which is cleaved by proteases in tissues in form "R," thereby releasing their ligands especially in the presence of inflammation. Although there is a great lack of knowledge on this subject, it is suspected that this very rare congenital deficiency may cause hypothyroidism [19,20].

REFERENCES

[1] Carrell R, Travis J. α1-Antitrypsin and the serpins: variation and countervariation. Trends Biochem Sci 1985;10:20–4.
[2] Heit C, Jackson BC, McAndrews M, Wright MW, Thompson DC, Silverman GA, et al. Update of the human and mouse SERPIN gene superfamily. Hum Genomics 2013;7:22.
[3] Silverman GA, Bird PI, Carrell RW, Church FC, Coughlin PB, Gettins PG, et al. The serpins are an expanding superfamily of structurally similar but functionally diverse proteins. Evolution, mechanism of inhibition, novel functions, and a revised nomenclature. J Biol Chem 2001;276:33293–6.
[4] Irving JA, Steenbakkers PJ, Lesk AM, Opden Camp HJ, Pike RN, Whisstock JC. Serpins in prokaryotes. Mol Biol Evol 2002;19:1881–90.
[5] Potempa J, Korzus E, Travis J. The serpin superfamily of proteinase inhibitors: structure, function, and regulation. J Biol Chem 1994;269:15957–60.
[6] van Gent D, Sharp P, Morgan K, Kalsheker N. Serpins: structure, function and molecular evolution. Int J Biochem Cell Biol 2003;35:1536–47.
[7] Gooptu B, Lomas DA. Conformational pathology of the serpins: themes, variations, and therapeutic strategies. Annu Rev Biochem 2009;78:147–76.
[8] Pike RN, Bottomley SP, Irving JA, Bird PI, Whisstock JC. Serpins: finely balanced conformational traps. IUBMB Life 2002;54:1–7.
[9] Marszal E, Shrake A. Serpin crystal structure and serpin polymer structure. Arch Biochem Biophys 2006;453:123–9.
[10] Ekeowa UI, Freeke J, Miranda E, Gooptu B, Bush MF, Pérez J, et al. Defining the mechanism of polymerization in the serpinopathies. Proc Natl Acad Sci USA 2010;107:17146–51.
[11] Kalsheker NA. Alpha 1-antichymotrypsin. Int J Biochem Cell Biol 1996;28:961–4.
[12] Janciauskiene S, Wright HT. Inflammation, antichymotrypsin, and lipid metabolism: autogenic etiology of Alzheimer's disease. Bioessays 1998;20:1039–46.
[13] Maclean PS, Tait RC. Hereditary and acquired antithrombin deficiency: epidemiology, pathogenesis and treatment options. Drugs 2007;67:1429–40.
[14] Davis AE 3rd. The pathophysiology of hereditary angioedema. Clin Immunol 2005;114:3–9.
[15] Christiansen SC, Zuraw BL. Update on therapeutic developments for hereditary angioedema. Allergy Asthma Proc 2009;30:500–5.
[16] Yepes M, Lawrence DA. Neuroserpin: a selective inhibitor of tissue-type plasminogen activator in the central nervous system. Thromb Haemost 2004;91:457–64.
[17] Zhou A, Wei Z, Stanley PL, Read RJ, Stein PE, Carrell RW. The S-to-R transition of corticosteroid-binding globulin and the mechanism of hormone release. J Mol Biol 2008;380:244–51.
[18] Lin HY, Muller YA, Hammond GL. Molecular and structural basis of steroid hormone binding and release from corticosteroid-binding globulin. Mol Cell Endocrinol 2010;316:3–12.
[19] Torpy DJ, Bachmann AW, Gartside M, Grice JE, Harris JM, Clifton P, et al. Association between chronic fatigue syndrome and the corticosteroid-binding globulin gene ALA SER224 polymorphism. Endocr Res 2004;30:417–29.
[20] Mannavola D, Vannucchi G, Fugazzola L, Cirello V, Campi I, Radetti G, et al. TBG deficiency: description of two novel mutations associated with complete TBG deficiency and review of the literature. J Mol Med (Berl) 2006;84:864–71.

Chapter 3

Alpha-1 Antitrypsin Biology

3.1 STRUCTURE

Human alpha-1 antitrypsin (AAT), also called alpha-1 proteinase inhibitor and SERPIN**A1** (serine protease inhibitor, clade **A**, member **1**), is a medium-sized glycoprotein (molecular weight 52 kDa), globular in shape (6.7 × 3.2 nm), water soluble, and diffusible into tissues, with an average half-life of around of 5 days in circulating blood [1]. The active site, formed by methionine and serine at position 358 and 359 respectively, protrudes from its globular contour (Fig. 3.1). The mature molecule is composed of a central chain of 394 amino acids and three branched carbohydrate side chains, attached to the protein in three asparagine (Asn) residues at positions 46, 83, and 247 (Fig. 3.2). Carbohydrates significantly contribute to prolong the molecule half-life, and give to it a marked electrophoretic heterogeneity [2].

Amino acids and carbohydrates are organized into 9 α-helixes (designated with the capital letters A–I) and 15 β-sheets designated A (*n*: 5), B (*n*: 6), and C (*n*: 4). Helixes and sheets form a compact frame holding a mobile loop of 20 residues, where the active site is located on exposed position to react with its substrates. In the trunk of the molecule, there are two cavities called "breach" and "shuter," in which many of the conformational physiological or mutational changes may occur [3] (Figs. 3.3 and 3.4).

3.2 SYNTHESIS AND SECRETION

Human AAT is primarily synthesized and secreted by hepatocytes (≈80%), and in additional quantities by monocytes, macrophages, alpha and delta cells of pancreas, type II alveolar epithelial lung cells, enterocytes, corneal keratinocytes and cancer cells from adenocarcinomas, sarcomas, glioblastomas, and cordomas [4–7]. In any case, hepatocyte expresses about 200 times more AAT messenger ribonucleic acid than any other cell [8].

At basal state, human body produces 34 mg/kg/day of AAT, and this activity is reflected by plasma concentrations of approximately 1–2 g/L. From plasma, 80%–90% of AAT diffuses to interstitial tissues, and up to 10% of this reaches biological fluids, including alveolar fluid (where local concentrations range

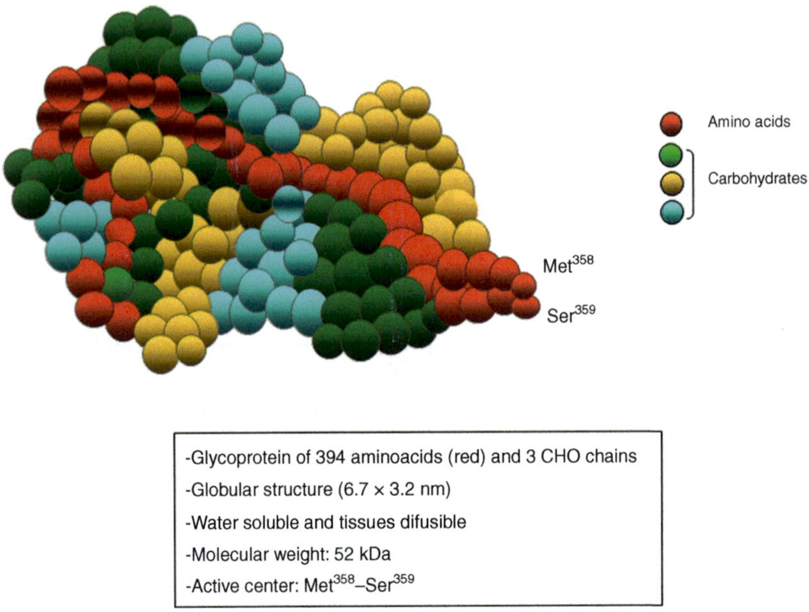

FIGURE 3.1 Surface representation of the alpha-1 antitrypsin molecule. *Red spheres* represent the amino acid chain, surrounded by other *green*, *blue*, and *yellow spheres* representing carbohydrates. Methionine and serine active site protrudes from the globular contour of the molecule. Adapted from de Serres F, Blanco I. Role of alpha-1 antitrypsin in human health and disease. J Intern Med 2014;276:311–35 [2].

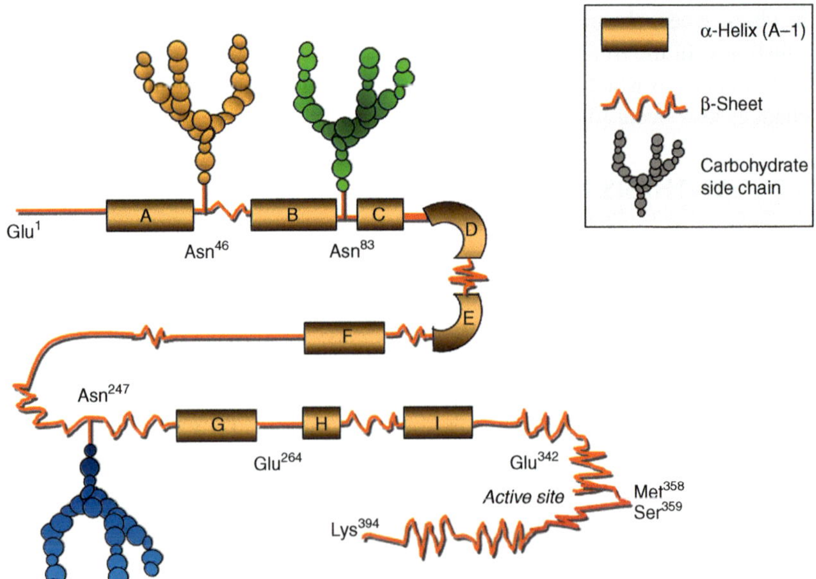

FIGURE 3.2 Schematic representation of the alpha-1 antitrypsin secondary structure. With protein chain unwound, the sheets and helices arrangements, the three carbohydrate side chains junction points, and the active site location in the polypeptide chain can be seen.

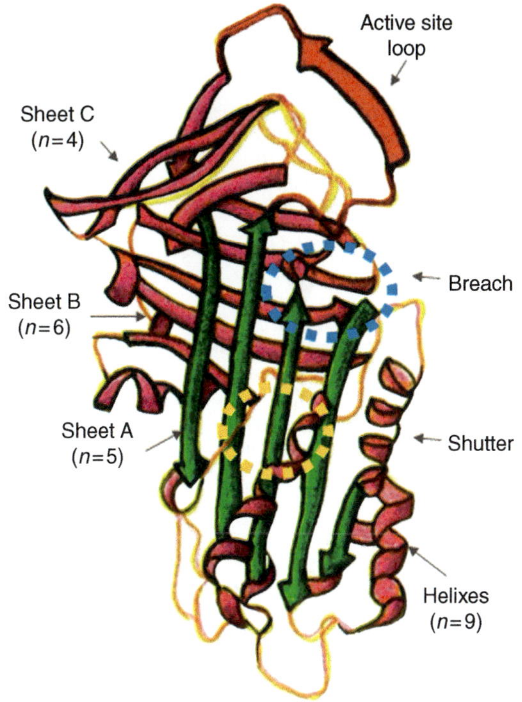

-3 groups of β-sheets (A–C)

-9 α-helixes

-A mobile loop of 20 aminoacids where the active site is exposed.

FIGURE 3.3 Tridimensional estructure of alpha-1 antitrypsin. Alpha-1antitrypsin is composed by three groups of β-sheets (A–C), nine α-helixes, and a mobile loop of 20 aminoacides where the active site is exposed. Two cavities called "breach" (circle of *blue dashes*) and "shutter" (circle of *yellow dashes*) are displayed.

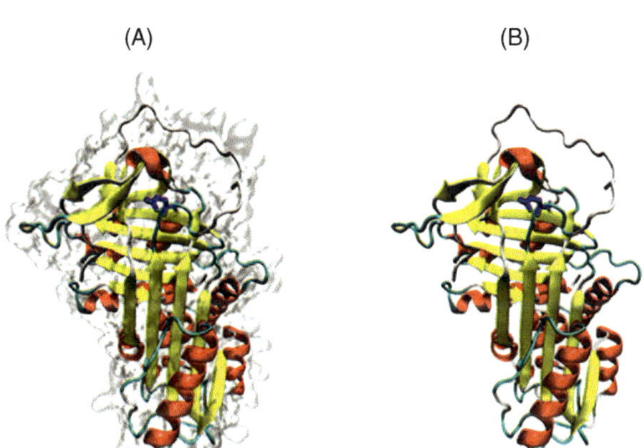

FIGURE 3.4 Graphical representation of covered (A) and noncovered (B) ribbon structure of the native alpha-1 antitrypsin molecule. Original figure adapted from the courtesy of Prof. Beatriz Martínez-Delgado, Instituto de Salud Carlos III, 28220 Majadahonda, Madrid, Spain.

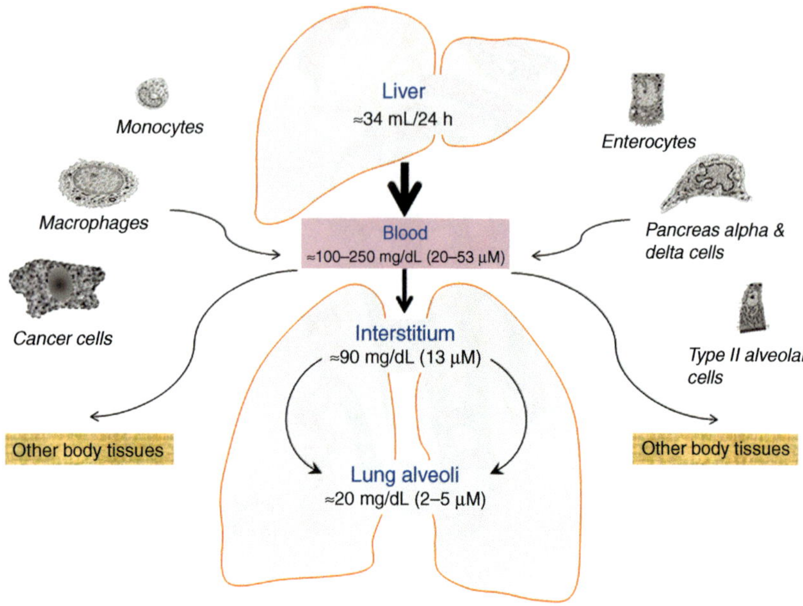

FIGURE 3.5 Sources and distribution of AAT in human body. Alpha-1 antitrypsin is mainly secreted by hepatocytes and in additional amounts by other cells. It reaches high blood concentrations, from which it diffuses to the interstices and body fluids. Its main function is to protect lungs (but also other tissues) from free proteases.

from 0.1 to 0.3 g/L) (Fig. 3.5), saliva, tears, milk, semen, bile, urine, and even cerebrospinal fluid (where it reaches relatively low concentrations of 0.006–0.02 g/L, which are markedly increased when meninges are inflamed) [9,10].

Recently, the Petrache's group has demonstrated that AAT is internalized by lung endothelial cells by clathrin- and caveolin-mediated endocytosis mechanisms [11], and this fact is likely extendable to other cell/tissues types expressing lipoprotein cell surface scavenger receptors class B, type I. This same group has also demonstrated the existence of an active transcellular transport of AAT macromolecules (i.e., transcytosis) between endothelial and lung epithelial cells [12].

As an acute phase reactant, AAT plasma levels increase rapidly up 2 times in response to inflammatory or infectious stimuli, accompanying C-reactive protein and serum amyloid A, and this increase gradually descends to baseline within 7 and 15 days. Serum levels are slightly increased by the use of contraceptives and during pregnancy [9].

In inflamed tissues the local concentration of AAT can rise up to 11 times, due to an increased synthesis by both resident cells and immigrants leukocytes in response to cytokines, leukotrienes, or endotoxins released by invading germs or inflamed tissues themselves. Local production of AAT is important to limit degradation of extracellular matrix by proteases released by activated cells [4,8,9].

3.3 SUBSTRATE

The specific substrate AAT is neutrophil elastase (synonyms: elastase 2 and ELA2), which reacts with one of the highest association constants known in biology ($k = 6.5 \times 10^7$ M/s) [13–15].

Neutrophil elastase is a serine protease stored in both primary (or azurophilic) granules and extracellular traps of neutrophils. It has multiple functions, including (1) intracellular degradation of proteins in lysosomes, (2) lysis of degraded products from inflamed tissues, (3) lysis of cell wall of invading bacteria, and (4) proteolysis of elastin and collagen-IV for extracellular matrix remodeling [16].

Elastase 2 gene consists of five exons and six introns, and its locus is located on chromosome 19 (19p13.3) adjacent to the proteinase-3 gene. ELA2 has a molecular weight of 29 kDa, and is formed by a glycosylated 218 amino acid chain and 2 lateral carbohydrate chains linked to the protein by 2 Asn residues. Its active site is formed by a catalytic triad of histidine–serine–asparagine (His41–Ser173–Asn88) [17] (Fig. 3.6).

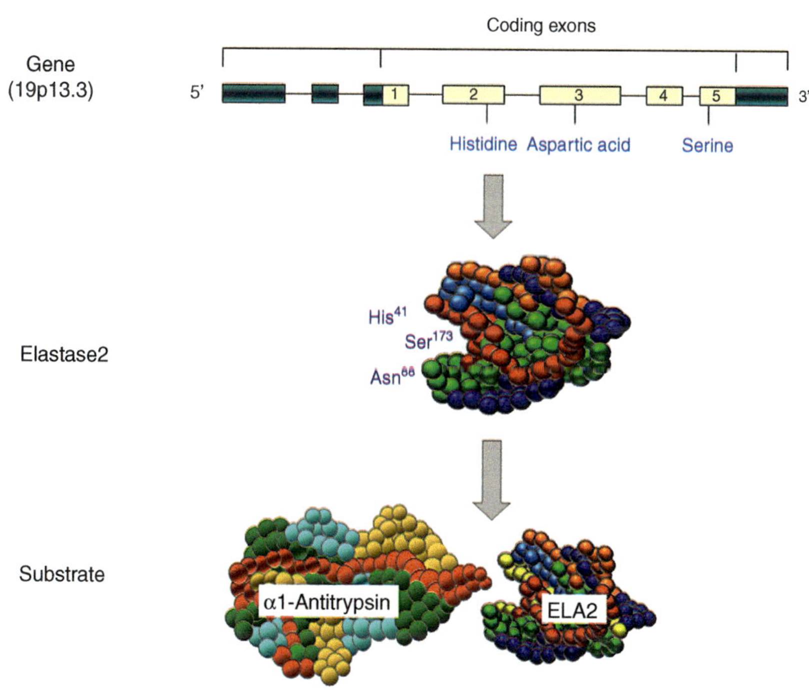

FIGURE 3.6 Neutrophil elastase: schematic representation of the gene and the protease cell surface. The neutrophil elastase gene has its locus in chromosome 19, and consists of 5 exons and 6 introns. It encodes neutrophil elastase or elastase 2 (ELA2), a 29-kDa glycoprotein with one chain of 218 amino acids and two carbohydrates linked-chains. Elastase is carried in azurophilic granules and extracellular traps of neutrophils. Its specific substrate is alpha-1 antitrypsin.

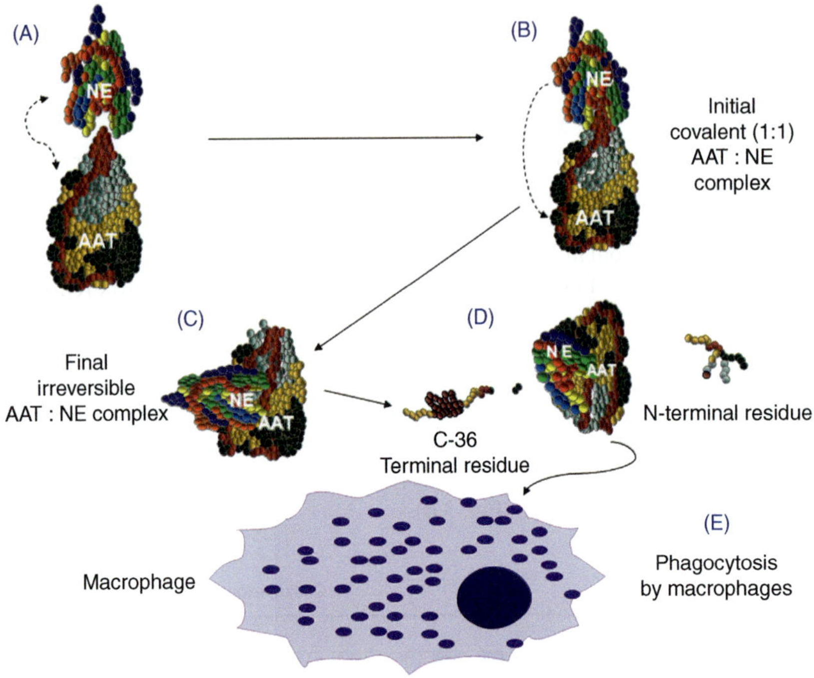

FIGURE 3.7 Alpha-1 antitrypsin (AAT) and neutrophil elastase (NE): inhibitory mechanisms. (A) The active site of AAT attracts neutrophil elastase, and a covalent stable complex between the two molecules is formed (B). Both molecules become permanently destructured, inactivated, and fused together. (C) The amino and carboxyl-terminal residues of the peptide chain break off (D), and the remaining complex is phagocytosed and cleared by macrophages (E).

The reaction mechanism between AAT and ELA2 is schematically illustrated in Fig. 3.7. Briefly, the AAT mobile loop acts as a "bait" attracting elastase. Elastase 2 (and other serine proteases) recognize the 20-amino acid "pseudosubstrate" within the AAT molecule that includes the reactive center loop. Upon binding, elastase cleaves AAT's reactive center loop, releasing stored potential energy and resulting in a conformational change in which the protease is catalytically inactivated. This action is followed by a dramatic conformational change in AAT, which captures elastase, and pulls it sharply from the top to the opposite pole of the molecule by a mechanism that has been compared to a "trap to catch mice" or a "catapult." In this collision, elastase is crushed and its catalytic triad inactivated. The final covalent (1:1) protease–antiprotease complex is the end result of an irreversible binding of both molecules, in which both become permanently destructured and inactivated in a process known as "suicide interaction." The AAT amino- and carboxyl-terminal residues are released, and the rest of the complex is captured by membrane receptors of macrophages, finally being internalized by endocytosis and degraded in lysosomas [4].

3.4 PROPERTIES

3.4.1 Inhibition of Extracellular Serine Proteases

Although the term "alpha-1-antitrypsin" is widely used by clinicians and researchers, it is a misnomer. This protein was first isolated in 1955 by Herman Schultze and named "alpha1-antitrypsin" because of its location in the alpha1-globulin fraction of the proteinogram and its ability to inhibit trypsin, and was used later when the AAT deficiency was discovered by Laurell and Eriksson in 1963 [9]. Over time, it became obvious that AAT is a major inhibitor of neutrophil elastase [13]. However, although the main function of AAT is inhibition of excessive free elastase from neutrophils, which reacts with an association constant 25 times greater than that applied to neutralize other proteases, several studies have shown that it can also effectively neutralize other serine proteinases from neutrophils, such as proteinase-3 [18,19], cathepsin-G [20], and defensin-α [21]; chymase and tryptase from mast cells [22–24]; trypsin, chymotrypsin, and elastase from pancreas [25,26]; granzyme B from T-lymphocytes [27]; and the circulating serine proteases plasmin, thrombin, plasminogen activator 1&2, factor Xa, and kallikreins 7&14 [28–32] (Fig. 3.8). These additional inhibitory capabilities are important because, for example, inhibition of thrombin and plasmin that reach an inflammatory focus after their extravasations from the

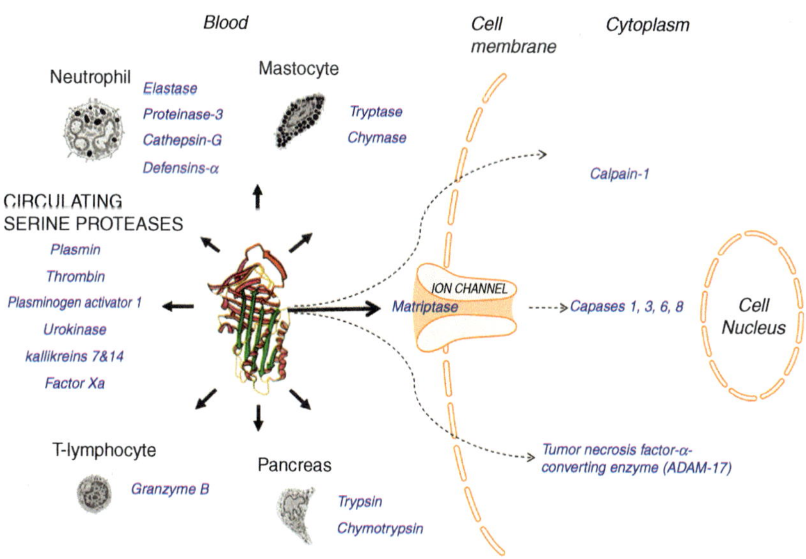

FIGURE 3.8 Alpha-1 antitrypsin antiprotease properties. Although the main substrate of alpha-1 antitrypsin is neutrophil elastase, its spectrum of action extends to other extracellular serine proteases, the cellular wall matriptase, and calpain 1, ADAM-17, and intracellular caspases.

blood through dilated blood vessels prevents activation of protease-activated receptors and several tissue proenzymes, such as metalloproteinases, kallikreins, or cathepsins, reducing inflammation intensity. These inhibitory properties of AAT provide to human plama more than 90% of its antiprotease capacity (the remaining 10% is provided by α2-macroglobulin) [2].

3.4.2 Inhibition of Transmembrane and Intracellular Proteases

Recent studies have shown that the antiprotease spectrum of AAT is not only limited to extracellular serine proteases, but also extends to other serine proteases of the cell membrane, and also to some intracellular cysteine proteases and metalloproteases (Fig. 3.8).

Specifically, AAT can deactivate the catalytic domain of matriptase slowly but steadily, and as this transmembrane-type serine proteinase is involved in the activation of lung epithelial sodium channels, AAT may contribute to modulate sodium transport and promote mucociliary clearance [33,34].

Several evidences indicate that AAT blocks apoptosis by binding and inhibiting the cysteine proteinase caspase-3, and modulates a broad range of inflammatory responses induced by lipopolysaccharide and tumor necrosis factor-alpha (TNF-α) from neutrophils [35]. For example, in cell-free studies and in a mouse model of apoptosis-dependent emphysema, it was shown that AAT is internalized by endothelial cells binding and inhibiting caspase-3, protecting against the alveolar enlargement caused by VEGF receptor blockade and by intratracheal instillation of caspase-3 [36].

In addition, strong antiapoptotic activities for AAT were observed when murine insulinoma cells (MIN6) were exposed to TNF-α. In a second model system of beta-cell apoptosis, treatment of MIN6 cells with AAT induced a significant increase in cellular viability and a reduction of apoptosis. Importantly, in both model systems, treatment with AAT completely abolished the induced caspase-3 activity [37].

In a mouse model of acute myocardial infarction, exogenous administration of AAT reduced caspase-1 activity in the ischemic myocardium leading to preservation of viable myocardium and prevention of adverse cardiac remodeling [38].

Recently, it has also been shown, in mouse models of Amanita phalloides α-amanitin and acetaminophen-induced liver injury, that administration of AAT prevented apoptosis and prolonged liver survival inhibiting capases 3 and 8 activation, and reducing serum levels and activity of TNF-α and ADAM17 metalloprotease (an acronym for "A Disintegrin and Metalloprotease"), also called TACE ("Tumor Necrosis Factor-A-Converting Enzyme"), which releases TNF-α attached to cell membranes by proteolysis of its anchors [39]. Reduced activity of ADAM17 by AAT had been previously discovered as a modulator of neutrophil chemotaxis [40]. More recently, it has also been found that AAT inactivates intracellularly neutrophil calpain I, a cysteine protease involved in motility and cell cycle, thereby reducing migration of leukocytes into tissues;

calpain-1 also processes the precursor of the proinflammatory interleukin-1α (IL-1α) expressed in virtually all epithelial cells [41].

3.4.3 Nonprotease-Inhibitory Antiinflammatory and Immunomodulatory Properties

Over the past 15 years, it has been shown that AAT has relevant antiinflammatory and immunomodulatory properties, modulating the activity of various key cells of the immune system, such as neutrophils, lymphocytes, monocytes, macrophages, and mast cells, as well as fibroblasts and epithelial cells [42].

Indeed, several studies have shown that AAT modulates neutrophil chemotaxis, degranulation, and apoptosis. For example, in vitro studies with peripheral blood neutrophils have demonstrated that AAT modulates neutrophil adhesion via binding and reduction of leukotriene B4 (LTB4) circulating levels [43,44]. In addition, AAT regulates human neutrophil chemotaxis induced by soluble immune complexes and interleukin-8 (IL-8) [40,45]. Other in vitro studies have demonstrated that AAT regulates neutrophil degranulation and autoimmunity, via binding two TNF-α types of receptors, that is, TNFR1 (TNF receptor type 1) and TNFR2 (TNF receptor type 2), modulating TNF-α receptor engagement, and thereby breaking the cycle of TNF-α-induced inflammation, preventing accelerated neutrophil apoptosis and improving bacterial killing [46].

In addition, a pilot study in humans has shown that short-term therapy with infusions of prolastin (60 mg/kg) at weekly intervals increased serum and lung secretion concentrations of AAT and was associated with a fall in LTB4, a potent cytokine chemoattractant of neutrophils central to the airway inflammation of AAT deficiency [47].

AAT also modulates cytokine expression in the blood by inhibiting synthesis and expression of the proinflammatory mediators IL-1, IL-6, IL-8, IL-32, MCP-1, and TNF-a, without interfering or even increasing the expression of the antiinflammatory products IL-1Ra, IL-10, and interferon gamma (IFNγ), downregulating hyperinflammation without impairing the normal immunoinflammatory response necessary to defend the body against infections/injuries [42,48,49].

In vitro and in vivo studies indicate that AAT is an endogenous inhibitor of nitric oxide (NO), a gas released by neutrophils, macrophages, and endothelial cells exposed to inflammatory stimuli, which regulates inflammation, vascular tone, neurotransmission, and immunity [50].

By its high content of methionine (nine residues), their ability to increase the activity of catalase (enzyme catalysing the decomposition of hydrogen peroxide—H_2O_2—into oxygen and water) and to reduce the neutrophil production of superoxide (O_2^-) AAT has a significant antioxidant power [51,52].

Of note, AAT nitrosylation by NO not only affects its inhibitory ability, but increases it against bacterial proteases [53].

However, in inflammatory media, oxidants and metalloproteinases produced by microbes and activated cells inactivate AAT by oxidation and cleavage of its active site, respectively [54,55].

At physiological concentrations, AAT contributes to the repair of the inflammation-damaged connective tissue by stimulating the fibroblasts production of procollagen and extracellular matrix [56].

3.4.4 Antimicrobial Properties

A growing number of studies have demonstrated in vitro and in vivo that a 20 residues fragment from the carboxy-terminal chain of AAT, called VIRIP (virus-inhibitory peptide) has a great ability to inhibit both replication and infectivity of the human immunodeficiency virus as well as other viruses, via binding and neutralizing the gp41 (glycoprotein 41) fusion peptide of the virus capsule, thus preventing it to entry into the cells [57–59].

In addition, it has been found that AAT has a significant bacteriostatic activity (possibly neutralizing bacterial serine proteases) against *Moraxella catarrhalis*, *Escherichia coli*, *Pseudomonas aeruginosa* [60–63], atypical mycobacteria [64], and the protozoa *Entamoeba histolytica* and *Cryptosporidium parvum* [65–67].

3.4.5 Protection and Repair of Damaged Tissues by Inflammation

AAT therapy was found to be beneficial in a wide range of experimental models. For example, in diabetic mice and pancreas transplanted rodents AAT increased the mass of beta cells protecting them from apoptosis, stimulated insulin production, and prolonged islet cell allografts from rejection [68–74]; AAT prevented graft-versus-host-disease [75–77]; delayed rheumatoid arthritis development [78]; protected from acute myocardial and renal ischemia–reperfusion injury [38]; inhibited acute liver failure [39]; and protected against TNF-α/endotoxin-induced lethality [79].

Moreover, apart from the demonstrated effectiveness of the AAT augmentation therapy to slow the progression of lung emphysema in patients with congenital AAT deficiency, this therapy has been found to be effective in single cases and small cohorts with clinical conditions other than lung emphysema, including panniculitis, vasculitis, fibromyalgia, and severe bronchial asthma [80–83]. In addition, several "in progress" clinical trials are showing a positive effect of AAT in type-1 diabetes (ClinicalTrials.gov Identifier: CT02093221, NCT01319331, NCT01304537, NCT01183468).

A summary of some AAT properties in vitro and in vivo is shown in Fig. 3.9.

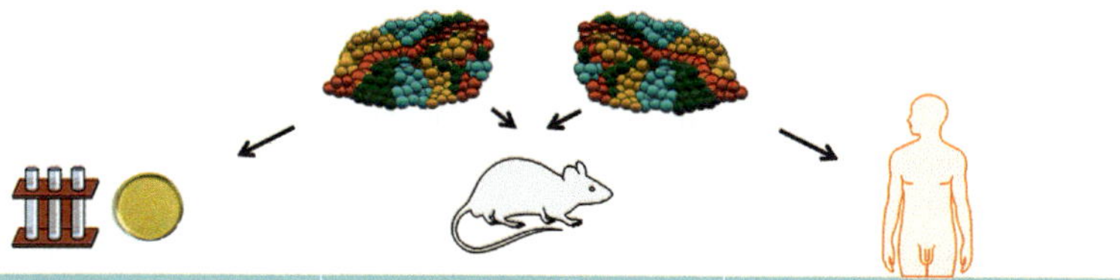

Blood tests and cell cultures	Animal models	Humans
Inhibits serine proteases (specially elastase), matryptase, caspase, calpain, and ADAM17	Protects against TNF-α and endotoxin-induced lethality	Reduces the loss of lung density in patients with AAT deficiency-related lung emphysema
Inhibits neutrophil and monocyte activation	Prevents lung emphysema development through inhibition of alveolar cells apoptosis	Reduces airway inflammation in patients with cystic fibrosis and AAT deficiency
Inhibits pancreas, lung, heart, vessels, and skin cells from apoptosis	Protects against acute lung, liver, heart, and renal ischemia–reperfusion injury	Controls clinical manifestations of AAT deficiency-associated systemic vasculitis
Inhibits LB-4, NO, TNF-α, IL-1β, IL-6, IL-8, IL-32, MCP-1 production without interfering with IL-10, IL-1Ra, or IFN-γ production	Ameliorates silica and cigarette smoke-induced emphysema	Controls clinical manifestations of AAT deficiency-associated neutrophilic panniculitis
Modulates activation, migration, and degranulation of circulating neutrofils	Increases the mass of pancreatic beta cells, and prolongs islets allograft survival	Controls symptoms of fibromyalgia in AAT-deficiency patients
Has antioxidant capacity	Prevents graft-versus-host-disease rejection	It has positive clinical effects in type-1 diabetes
Inhibits replication and infectivity of HIV, and other viruses, bacteria, and protozoa	Decreases autoimmunity and delays rheumatoid arthritis development	
	Protects and repairs connective tissue damaged by inflammation	

FIGURE 3.9 Biological activities of alpha-1 antitrypsin "in vitro" and "in vivo." Alpha-1 antitrypsin is a multifunctional glycoprotein which provides to human organisms antiinflammatory immunomodulatory, antimicrobial, and tissue repair capacities. *IL*, Interleukin; *IL-1Ra*, interleukin-1 receptor antagonist; *LB-4*, leukotriene B4; *TNF*, tumor necrosis factor; *HIV*, human immunodeficiency virus.

REFERENCES

[1] American Thoracic Society/European Respiratory Society. American Thoracic Society/European Respiratory Society Statement: standards for the diagnosis and management of individuals with alpha1-antitrypsin deficiency. Am J Respir Crit Care Med 2003;168:818–900.

[2] de Serres F, Blanco I. Role of alpha-1 antitrypsin in human health and disease. J Intern Med 2014;276:311–35.

[3] Lomas DA, Parfrey H. Alpha1-antitrypsin deficiency. 4: molecular pathophysiology. Thorax 2004;59:529–35.

[4] Stoller JK, Aboussouan LS. Alpha1-antitrypsin deficiency. Lancet 2005;365:2225–36.

[5] Mornex JF, Chytil-Weir A, Martinet Y, Courtney M, LeCocq JP, Crystal RG. Expression of the al-antitrypsin gene in mononuclear phagocytes of normal and a1-antitrypsin-deficient individuals. J Clin Invest 1986;77:1952–61.

[6] Perlmutter DH, Cole FS, Kilbridge P, Rossing TH, Colten HR. Expression of a1-proteinase inhibitor gene in human monocytes and macrophages. Proc Natl Acad Sci USA 1985;82:795–9.

[7] Pérez-Holanda S, Blanco I, Menéndez M, Rodrigo L. Serum concentration of alpha-1 antitrypsin is significantly higher in colorectal cancer patients than in healthy controls. BMC Cancer 2014;14:355.

[8] McCarthy C, Saldova R, Wormald MR, Rudd PM, McElvaney NG, Reeves EP. The role and importance of glycosylation of acute phase proteins with focus on alpha-1 antitrypsin in acute and chronic inflammatory conditions. J Proteome Res 2014;13:3131–43.

[9] Janciauskiene SM, Bals R, Koczulla R, Vogelmeier C, Köhnlein T, Welte T. The discovery of α1-antitrypsin and its role in health and disease. Respir Med 2011;105:1129–39.
[10] Chen CP1, Chen RL, Preston JE. The influence of ageing in the cerebrospinal fluid concentrations of proteins that are derived from the choroid plexus, brain, and plasma. Exp Gerontol 2012;47:323–8.
[11] Lockett AD, Petrusca DN, Justice MJ, Poirier C, Serban KA, Rush NI, et al. Scavenger receptor class B, type I mediated uptake of A1AT by pulmonary endothelial c. Am J Physiol Lung Cell Mol Physiol 2015;309:L425–34.
[12] Lockett AD, Brown MB, Santos-Falcon N, Rush NI, Oueini H, Oberle AJ, et al. Active trafficking of alpha 1 antitrypsin across the lung endothelium. PLoS One 2014;9:e93979.
[13] Janoff A. Inhibition of human granulocyte elastase by serum alpha-1-antitrypsin. Am Rev Respir Dis 1972;105:121–2.
[14] Sinden NJ, Baker MJ, Smith DJ, et al. α-1-antitrypsin variants and the proteinase/antiproteinase imbalance in chronic obstructive pulmonary disease. Am J Physiol Lung Cell Mol Physiol 2015;308:L179–90.
[15] Stoller JK, Lacbawan FL, Aboussouan LS. Alpha-1 antitrypsin deficiency. In: Pagon RA, Adam MP, Ardinger HH, Bird TD, Dolan CR, Fong CT, Smith RJH, Stephens K, editors. GeneReviews® [Internet]. Seattle, WA: University of Washington; 1993–2014.
[16] Travis J. Structure, function, and control of neutrophil proteinases. Am J Med 1988;84:37–42.
[17] Takahashi H1, Nukiwa T, Yoshimura K, Quick CD, States DJ, Holmes MD, et al. Structure of the human neutrophil elastase gene. J Biol Chem 1988;263:14739–47.
[18] Travis J, Salvesen GS. Human plasma proteinase inhibitors. Annu Rev Biochem 1983;52: 655–709.
[19] Duranton J, Bieth JG. Inhibition of proteinase 3 by [alpha]1-antitrypsin in vitro predicts very fast inhibition in vivo. Am J Respir Cell Mol Biol 2003;29:57–61.
[20] Duranton J, Adam C, Bieth JG. Kinetic mechanism of the inhibition of cathepsin G by alpha 1-antichymotrypsin and alpha 1-proteinase inhibitor. Biochemistry 1998;37:11239–45.
[21] Spencer LT, Paone G, Krein PM, Rouhani FN, Rivera-Nieves J, Brantly ML. Role of human neutrophil peptides in lung inflammation associated with alpha1-antitrypsin deficiency. Am J Physiol Lung Cell Mol Physiol 2004;286:L514–20.
[22] He SH, Chen HQ, Zheng J. Inhibition of tryptase and chymase induced nucleated cell infiltration by proteinase inhibitors. Acta Pharmacol Sin 2004;25:1677–84.
[23] He SH, Xie H, Zhang XJ, Wang XJ. Inhibition of histamine release from human mast cells by natural chymase inhibitors. Acta Pharmacol Sin 2004;25:822–6.
[24] He SH, Xie H. Inhibition of tryptase release from human colon mast cells by protease inhibitors. World J Gastroenterol 2004;10:332–6.
[25] Rowley PT, Sevilla ML, Schwartz H. Serum alpha-antitrypsin types: elastase inhibition versus trypsin inhibition. Hum Hered 1974;24:472–81.
[26] Cohen AB. The interaction of alpha-1-antitrypsin with chymotrypsin, trypsin and elastase. Biochim Biophys Acta 1975;391:193–200.
[27] Poe M, Blake JT, Boulton DA, Gammon M, Sigal NH, Wu JK, et al. Human cytotoxic lymphocyte granzyme B. Its purification from granules and the characterization of substrate and inhibitor specificity. J Biol Chem 1991;266:98–103.
[28] Gallimore MJ. Serum inhibitors in fibrinolysis. Br J Haematol 1975;31:217–31.
[29] Clemmensen I, Christensen F. Inhibition of urokinase by complex formation with human alpha1-antitrypsin. Biochim Biophys Acta 1976;429:591–9.
[30] Gitel SN, Medina VM, Wessler S. Inhibition of human activated factor X by antithrombin III and alpha 1-proteinase inhibitor in human plasma. J Biol Chem 1984;259:6890–5.

[31] Patston PA, Roodi N, Schifferli JA, Bischoff R, Courtney M, Schapira M. Reactivity of alpha 1-antitrypsin mutants against proteolytic enzymes of the kallikrein-kinin, complement, and fibrinolytic systems. J Biol Chem 1990;265:10786–91.

[32] Luo LY, Jiang W. Inhibition profiles of human tissue kallikreins by serine protease inhibitors. Biol Chem 2006;387:813–6.

[33] Janciauskiene S, Nita I, Subramaniyam D, Li Q, Lancaster JR Jr, Matalon S. Alpha1-antitrypsin inhibits the activity of the matriptase catalytic domain in vitro. Am J Respir Cell Mol Biol 2008;39:631–7.

[34] Lazrak A, Nita I, Subramaniyam D, Wei S, Song W, Ji HL, et al. Alpha(1)-antitrypsin inhibits epithelial Na+ transport in vitro and in vivo. Am J Respir Cell Mol Biol 2009;41:261–70.

[35] Tuder RM, Janciauskiene SM, Petrache I. Lung disease associated with alpha1-antitrypsin deficiency. Proc Am Thorac Soc 2010;7:381–6.

[36] Petrache I, Fijalkowska I, Medler TR, Skirball J, Cruz P, Zhen L, et al. Alpha-1 antitrypsin inhibits caspase-3 activity, preventing lung endothelial cell apoptosis. Am J Pathol 2006;169:1155–66.

[37] Zhang B, Lu Y, Campbell-Thompson M, Spencer T, Wasserfall C, Atkinson M, et al. Alpha1-antitrypsin protects beta-cells from apoptosis. Diabetes 2007;56:1316–23.

[38] Toldo S, Seropian IM, Mezzaroma E, Van Tassell BW, Salloum FN, Lewis EC, et al. Alpha-1 antitrypsin inhibits caspase-1 and protects from acute myocardial ischemia-reperfusion injury. J Mol Cell Cardiol 2011;51:244–51.

[39] Jedicke N, Struever N, Aggrawal N, Welte T, Manns MP, Malek NP, et al. Alpha-1-antitrypsin inhibits acute liver failure in mice. Hepatology 2014;59:2299–308.

[40] Bergin DA, Reeves EP, Meleady P, Henry M, McElvaney OJ, Carroll TP, et al. α-1 Antitrypsin regulates human neutrophil chemotaxis induced by soluble immune complexes and IL-8. J Clin Invest 2010;120:4236–50.

[41] Al-Omari M, Korenbaum E, Ballmaier M, Lehmann U, Jonigk D, Manstein DJ, et al. The acute phase protein, α1-antitrypsin, inhibits neutrophil calpain I and induces random migration. Mol Med 2011;17:865–74.

[42] Lewis EC. Expanding the clinical indications for α(1)-antitrypsin therapy. Mol Med 2012;18:957–70.

[43] Hurley K, Lacey N, O'Dwyer CA, Bergin DA, McElvaney OJ, O'Brien ME, et al. Alpha-1 antitrypsin augmentation therapy corrects accelerated neutrophil apoptosis in deficient individuals. J Immunol 2014;193:3978–91.

[44] O'Dwyer CA, O'Brien ME, Wormald MR, White MM, Banville N, Hurley K, et al. The BLT1 inhibitory function of α-1 antitrypsin augmentation therapy disrupts leukotriene B4 neutrophil signaling. J Immunol 2015;195:3628–41.

[45] Hubbard RC, Fells G, Gadek J, Pacholok S, Humes J, Crystal RG. Neutrophil accumulation in the lung in alpha 1-antitrypsin deficiency. Spontaneous release of leukotriene B4 by alveolar macrophages. J Clin Invest 1991;88:891–7.

[46] Bergin DA, Reeves EP, Hurley K, Wolfe R, Jameel R, Fitzgerald S, et al. The circulating proteinase inhibitor α-1 antitrypsin regulates neutrophil degranulation and autoimmunity. Sci Transl Med 2014;6(217). 217ra1.

[47] Stockley RA, Bayley DL, Unsal I, Dowson LJ. The effect of augmentation therapy on bronchial inflammation in alpha1-antitrypsin deficiency. Am J Respir Crit Care Med 2002;165:1494–8.

[48] Pott GB, Chan ED, Dinarello CA, Shapiro L. Alpha-1-antitrypsin is an endogenous inhibitor of proinflammatory cytokine production in whole blood. J Leukoc Biol 2009;85:886–95.

[49] Subramaniyam D, Virtala R, Pawłowski K, Clausen IG, Warkentin S, Stevens T, et al. TNF-alpha-induced self-expression in human lung endothelial cells is inhibited by native and oxidized alpha1-antitrypsin. Int J Biochem Cell Biol 2008;40:258–71.

[50] Chan ED, Pott GB, Silkoff PE, Ralston AH, Bryan CL, Shapiro L. Alpha-1-antitrypsin inhibits nitric oxide production. J Leukoc Biol 2012;92:1251–60.

[51] Levine RL, Berlett BS, Moskovitz J, Mosoni L, Stadtman ER. Methionine residues may protect proteins from critical oxidative damage. Mech Ageing Dev 1999;107:323–32.

[52] Bucurenci N, Blake DR, Chidwick K, Winyard PG. Inhibition of neutrophil superoxide production by human plasma alpha 1-antitrypsin. FEBS Lett 1992;300:21–4.

[53] Miyamoto Y, Akaike T, Alam MS, Inoue K, Hamamoto T, Ikebe N, et al. Novel functions of human alpha(1)-protease inhibitor after S-nitrosylation: inhibition of cysteine protease and antibacterial activity. Biochem Biophys Res Commun 2000;267:918–23.

[54] Taggart C, Cervantes-Laurean D, Kim G, McElvaney NG, Wehr N, Moss J, et al. Oxidation of either methionine 351 or methionine 358 in alpha 1-antitrypsin causes loss of anti-neutrophil elastase activity. J Biol Chem 2000;275:27258–65.

[55] Liu Z, Zhou X, Shapiro SD, Shipley JM, Twining SS, Diaz LA, et al. The serpin alpha1-proteinase inhibitor is a critical substrate for gelatinase B/MMP-9 in vivo. Cell 2000;102:647–55.

[56] Dabbagh K, Laurent GJ, Shock A, Leoni P, Papakrivopoulou J, Chambers RC. Alpha-1-antitrypsin stimulates fibroblast proliferation and procollagen production and activates classical MAP kinase signalling pathways. J Cell Physiol 2001;186:73–5.

[57] Shapiro L, Pott GB, Ralston AH. Alpha-1-antitrypsin inhibits human immunodeficiency virus type 1. FASEB J 2001;15:115–22.

[58] Congote LF. The C-terminal 26-residue peptide of serpin A1 is an inhibitor of HIV-1. Biochem Biophys Res Commun 2006;343:617–22.

[59] Bryan CL, Beard KS, Pott GB, Rahkola J, Gardner EM, Janoff EN, et al. HIV infection is associated with reduced serum alpha-1-antitrypsin concentrations. Clin Invest Med 2010;33:E384–9.

[60] Hadzic R, Nita I, Tassidis H, Riesbeck K, Wingren AG, Janciauskiene S. Alpha1-antitrypsin inhibits Moraxella catarrhalis MID protein-induced tonsillar B cell proliferation and IL-6 release. Immunol Lett 2006;102:141–7.

[61] Knappstein S, Ide T, Schmidt MA, Heusipp G. Alpha 1-antitrypsin binds to and interferes with functionality of EspB from atypical and typical enteropathogenic Escherichia coli strains. Infect Immun 2004;72:4344–50.

[62] Cantin AM, Woods DE. Aerosolized prolastin suppresses bacterial proliferation in a model of chronic Pseudomonas aeruginosa lung infection. Am J Respir Crit Care Med 1999;160: 1130–5.

[63] Kaner Z, Ochayon DE, Shahaf G, Baranovski BM, Bahar N, Mizrahi M, et al. Acute phase protein α1-antitrypsin reduces bacterial burden in mice by selective modulation of innate cell responses. J Infect Dis 2015;211:1489–98.

[64] Chan ED, Kaminska AM, Gill W, Chmura K, Feldman NE, Bai X, et al. Alpha-1-antitrypsin (AAT) anomalies are associated with lung disease due to rapidly growing mycobacteria and AAT inhibits Mycobacterium abscessus infection of macrophages. Scand J Infect Dis 2007;39:690–6.

[65] Lushbaugh WB, Kairalla AB, Hofbauer AF, Arnaud P, Cantey JR, Pittman FE. Inhibition of Entamoeba histolytica cytotoxin by alpha 1 antiprotease and alpha 2 macroglobulin. Am J Trop Med Hyg 1981;30:575–85.

[66] Forney JR, Yang S, Healey MC. Antagonistic effect of human alpha-1-antitrypsin on excystation of Cryptosporidium parvum oocysts. J Parasitol 1997;83:771–4.

[67] Forney JR, Yang S, Healey MC. Synergistic anticryptosporidial potential of the combination alpha-1-antitrypsin and paromomycin. Antimicrob Agents Chemother 1997;41:2006–8.

[68] Koulmanda M, Bhasin M, Hoffman L, Fan Z, Qipo A, Shi H, et al. Curative and beta cell regenerative effects of alpha1-antitrypsin treatment in autoimmune diabetic NOD mice. Proc Natl Acad Sci USA 2008;105:16242–7.

[69] Loganathan G, Dawra RK, Pugazhenthi S, Wiseman AC, Sanders MA, Saluja AK, et al. Culture of Impure Human Islet Fractions in the Presence of Alpha-1 Antitrypsin Prevents Insulin Cleavage and Improves Islet Recovery. Transplant Proc 2010;42:2055–7.

[70] Kalis M, Kumar R, Janciauskiene S, Salehi A, Cilio CM. Alpha-1 antitrypsin enhances insulin secretion and prevents cytokine-mediated apoptosis in pancreatic b-cells. Islets 2010;2:185–9.

[71] Shahaf G, Moser H, Ozeri E, Mizrahi M, Abecassis A, Lewis EC. α-1-antitrypsin gene delivery reduces inflammation, increases T-regulatory cell population size and prevents islet allograft rejection. Mol Med 2011;17:1000–11.

[72] Ashkenazi E, Baranovski BM, Shahaf G, Lewis EC. Pancreatic islet xenograft survival in mice is extended by a combination of alpha-1-antitrypsin and single-dose anti-CD4/CD8 therapy. PLoS One 2013;8:e63625.

[73] Bellacen K, Kalay N, Ozeri E, Shahaf G, Lewis EC. Revascularization of pancreatic islet allografts is enhanced by α-1-antitrypsin under anti-inflammatory conditions. Cell Transplant 2013;22:2119–33.

[74] Yang L, Liao YT, Yang XF, Reng LW, Qi H, Li FR. Immune protective effect of human alpha-1-antitrypsin gene during β cell transplantation in diabetic mice. Immunol Res 2015; 62:71–80.

[75] Marcondes AM, Xiang L, Tabellini L, Bartenstein M, Kabacka J, Sale GE, et al. Inhibition of IL-32 activation by a-1 anti-trypsin reduces alloreactivity and increases survival in an allogeneic murine marrow transplantation model. Blood 2011;118:5031–9.

[76] Tawara I, Sun Y, Lewis EC, Toubai T, Evers R, Nieves E, et al. Alpha-1-antitrypsin monotherapy reduces graft-versus-host disease after experimental allogeneic bone marrow transplantation. Proc Natl Acad Sci USA 2012;109:564–9.

[77] Mizrahi M, Cal P, Rosenthal M, Ochayon D, Shahaf G, Kaner Z, et al. Human α1-antitrypsin modifies B-lymphocyte responses during allograft transplantation. Immunology 2013;140:362–73.

[78] Grimstein C, Choi YK, Wasserfall CH, Satoh M, Atkinson MA, Brantly ML, et al. Alpha-1 antitrypsin protein and gene therapies decrease autoimmunity and delay arthritis development in mouse model. J Transl Med 2011;9:21.

[79] Libert C, Van Molle W, Brouckaert P, Fiers W. alpha1-Antitrypsin inhibits the lethal response to TNF in mice. J Immunol 1996;157:5126–9.

[80] Blanco I, Lara B, de Serres F. Efficacy of alpha1-antitrypsin augmentation therapy in conditions other than pulmonary emphysema. Orphanet J Rare Dis 2011;6:14.

[81] Blanco I, Lipsker D, Lara B, Janciauskiene S. Neutrophilic panniculitis associated with alpha-1 antitrypsin deficiency: an update. Br J Dermatol 2016;174:753–62.

[82] Blanco I, Canto H, Flores J, Camblor C, Carcaba V, de Serres FJ, et al. Long-term augmentation therapy with alpha-1 antitrypsin in an MZ-AAT severe persistent asthma. Monaldi Arch Chest Dis 2008;69:178–82.

[83] Blanco I, Canto H, de Serres FJ, Fernandez-Bustillo E, Rodríguez MC. Alpha1-antitrypsin replacement therapy controls fibromyalgia symptoms in 2 patients with PI ZZ alpha1-antitrypsin deficiency. J Rheumatol 2004;31:2082–5.

Chapter 4

Alpha-1 Antitrypsin Gene, Genetic Heritage, Phenotypes, and Genotypes

4.1 THE ALPHA-1 ANTITRYPSIN GENE

4.1.1 Locus

The alpha-1 antitrypsin (AAT) gene's official symbol is SERPINA1 [serpin peptidase inhibitor, clade A (alpha-1 antiproteinase, antitrypsin), member 1]. It is located at the distal end of the long arm (q) of the 14th chromosome, where it occupies a short space of 12.2 kb at position q31-32.3 (base pairs: 94,376,747–94,390,692). SERPINA1 shares locus with other genes encoding other serpins, such as SERPINA3 (α1-antichymotrypsin), SERPINA4 (kallistatin), SERPINA5 (protein C inhibitor), SERPINA6 (cortisol-binding globulin, or transcortin), SERPINA9 (centerin), SERPINA10 [protein Z-dependent protease inhibitor (Pi)], SERPINA11 (antiproteinase-like 2 serine or cysteine proteinase inhibitor, clade A, member 11), and a "dumb" pseudogene with an obscure meaning [1,2] (Fig. 4.1).

4.1.2 Expression Regulation

SERPINA1 gene consists of seven exons (designated IA, IB, IC, II, III, IV, and V) and six interspersed introns. Gene expression is, predominantly, directed by two regulatory elements from the promoter, untranslated region 5′UTR: (1) a specific promoter for hepatocytes with transcription initiation in exon IC, and (2) two other alternative promoter regions for monocytes and macrophages located in exons IA and IB [3–6].

Antitrypsin is initially an immature protein of 418 amino acids, which once secreted in its mature form has 394 amino acids after removing a 24-amino acid signaling peptide [3]. The translation sites are located between the coding exons II and V. The active site is encoded in exon V. The asparagine (Asn) molecules to which three carbohydrate chains are linked are encoded in exons II and III. In the distal part of the gene is the TAA (thymine, adenine, adenine) nucleotide triplet stop codon that signals the translation termination (Fig. 4.1).

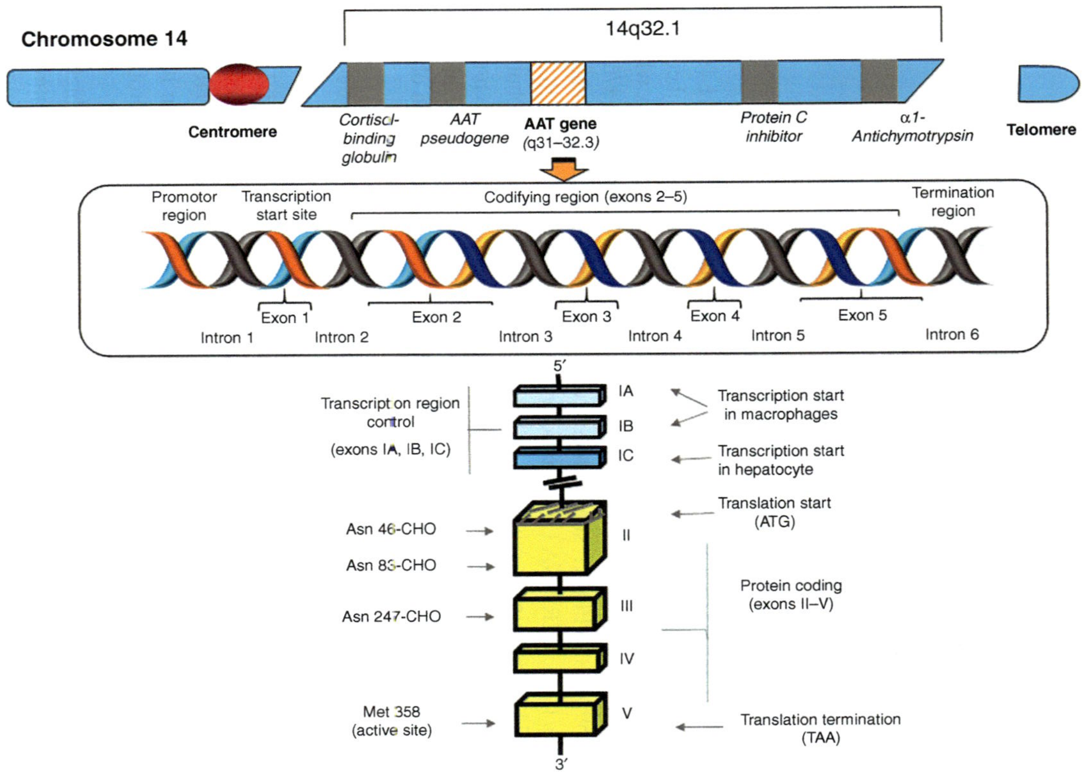

FIGURE 4.1 **Gene locus and schematic structure of the alpha-1 antitrypsin (SERPINA1) gene.** At the top of the figure, is represented the human alpha-1 antitrypsin gene locus on chromosome 14q32.1, along with other serpin genes forming a cluster at this locus. At the bottom, a diagram of the gene with various points marked with arrows, where the sites of various activities of protein synthesis are located.

4.1.3 Transcription Stimuli

SERPINA1 is fundamentally an antiinflammatory gene primarily activated by products generated during inflammatory processes. The best known stimuli are not only the bacterial endotoxin or lipopolysaccharide, but also interferon-β, IL-1, IL-6, and TNF-alpha cytokines, as well as various products resulting from nitrooxidative stress which are generically called *advanced glycation end products* (AGEs) activate this gene [6]. Other gene activation mechanisms that function as self-regulatory of the protease–antiprotease balance involve circulating serine proteases, AAT-elastase complexes, and carboxy terminal peptide fragments detached from the AAT: elastase complex after being hydrolyzed [7]. Likewise, the gene can be stimulated by sex steroid hormones [8].

All of these activating agents stimulate several intracellular transcription pathways acting previously on different surface receptors, such as toll-like receptors for lipopolysaccharide, several types of G protein-coupled receptors for cytokines, immunoglobulin-type receptors for AGEs and lipoprotein-like receptors for serine proteases, and AAT-elastase complexes [9–11]. Estrogens, androgens, and anabolic steroids could modulate AAT biosynthesis crossing through the cell membrane and binding to specific cytosolic hormone receptors that dimerize and translocate to the nucleus to activate specific hormone response element sequences of DNA (Fig. 4.2).

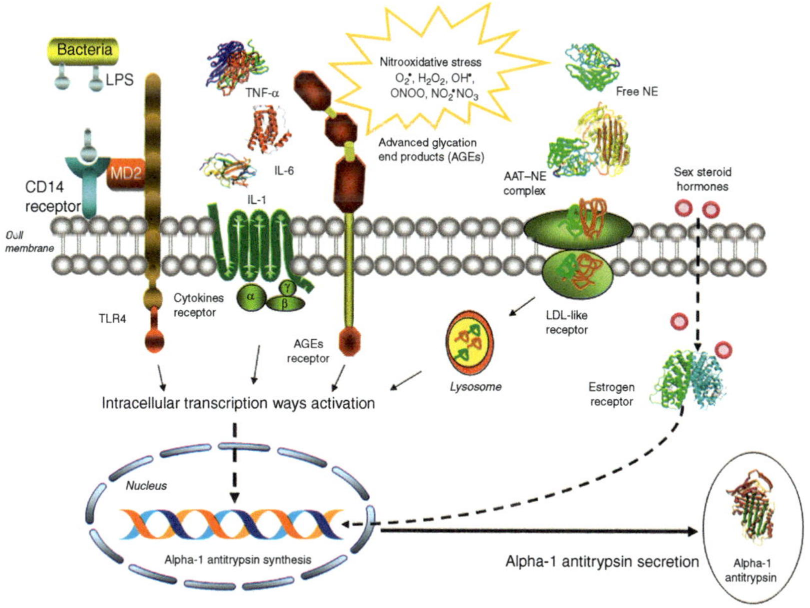

FIGURE 4.2 **Transcription stimuli for the SERPINA1 gene.** *AGE*, Advanced glycation end products; *CD14*, cluster of differentiation 14; *IL*, interleukin; *LDL*, low-density lipoprotein; *LPS*, lipopolysaccharide; *MD2*, protein MD2 (lymphocyte antigen 96); *NE*, neutrophil elastase; *TLR*, toll-like receptor; *TNF*, tumor necrosis factor.

4.1.4 AAT Transcription

These intracellular signaling cascades interact in the promoter region of the gene with various transcription factors, including the TATA (thymine–adenine–thymine–adenine) box, and then with the RNA polymerase (or transcriptase), an enzyme that replicates DNA to synthesize RNA from a DNA sequence serving as pattern or mold. This initial messenger RNA (mRNA) is called primary mRNA transcript, as well as pre-mRNA [12]. Next pre-mRNA undergoes a process of maturation, which basically involves a removal of all introns (*splicing*) by five small nuclear ribonucleoproteins. Finally, this mature mRNA diffuses into the rough endoplasmic reticulum (RER) carrying the genetic message to the translational apparatus of ribosomes for its translation (Fig. 4.3).

4.1.5 Intracellular Mechanisms of Protein Folding and Secretion

Ribosomes link amino acids together in the order specified by the mature mRNA. The small ribosomal subunit reads the mRNA, and the large subunit

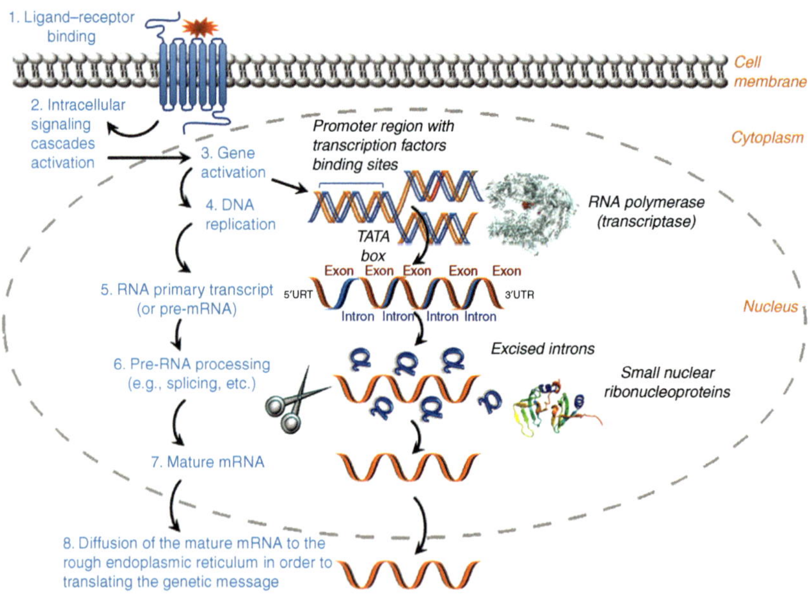

FIGURE 4.3 Alpha-1 antitrypsin transcription, DNA replication, and messenger RNA processing. After binding of ligands with receptors, intracellular signaling cascades are activated and then interacting in the promoter region of the AAT gene with various transcription factors and RNA polymerase (or activated transcriptase) that replicates DNA for synthesizing a RNA type called primary transcript or pre-RNA. Pre-RNA undergoes a maturation process that includes a removing of all introns by five small nuclear ribonucleoproteins (snRNP). Finally, the mature messenger RNA (mRNA) passes trough the rough endoplasmic reticulum to take part in the translation of the genetic message.

joins the amino acids carried by a transfer RNA to form a polypeptide chain. These chains of immature polypeptide molecules are deposited in the RER, and then recognized by the "quality control apparatus" of the cell, essentially consisting of mannosidase I and a group of proteins generically called chaperones, the most important calnexin and ubiquitin.

Chaperones bind to the newly formed chains of polypeptides to provide to these molecules a correct folded conformation. Folded molecules pass to the Golgi apparatus where they are glycosylated, purified, and packaged into secretory vesicles, from which they are finally secreted in the bloodstream. Misfolded defective molecules are detected by mannosidase 1 and marked by ubiquitin and calnexin to form trimolecular (ubiquitin–AAT–calnexin) complexes which are expelled from the RER and directed to the "crematorium barrel" of the proteasome, where they are degraded, reduced to short peptide chains or isolated amino acids, which can be reused by the same cell [13] (Fig. 4.4).

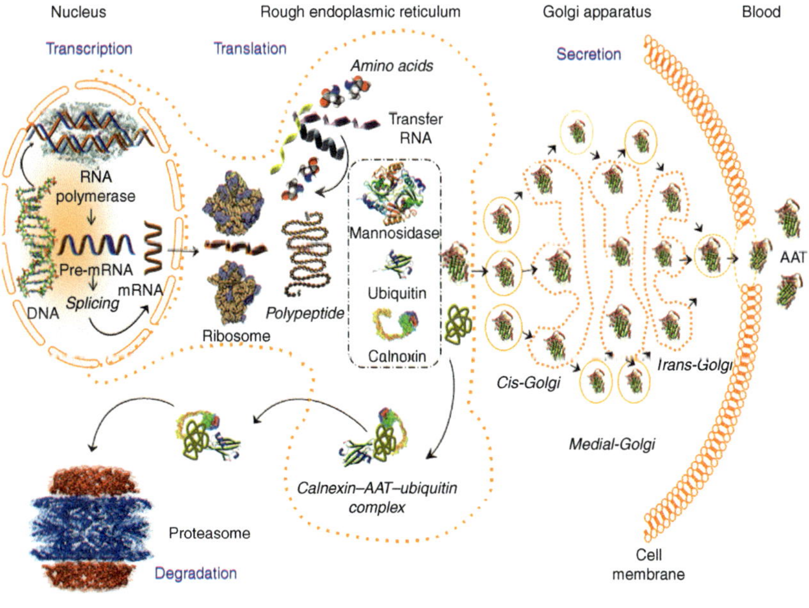

FIGURE 4.4 Transcription, translation, degradation, and secretion of alpha-1 antitrypsin. After the transcription processes, mature mRNA carries the genetic message to ribosomes, where the translation and orderly assembly of amino acids carried by a transfer RNA (tRNA) occurs to form polypeptide chains. These new immature protein molecules are recognized by the "quality control apparatus" of the RER (i.e., mannosidase-1, calnexin, and ubiquitin) that help to provide to the immature molecules a correctly folded conformation. Correctly folded molecules pass through the Golgi and are secreted into the blood. Misfolded molecules are detected by mannosidase 1 and marked by ubiquitin and calnexin to be discarded and sent to the proteasome for proteolysis.

4.2 THE "PROTEASE INHIBITOR" SYSTEM

In 1967, Magne Fagerhol and Carl-Bertil Laurell reported that the AAT had a polymorphic alpha-1 band with several different patterns in starch gel and acidic pH, and proposed to call "Pi" system to the set of the AAT electrophoretic variants [14]. Initially, they designated the AAT variants with the alphabet letters according to their electrophoretic migration velocity in starch gels (a technique that was later abandoned). They called Pi M (from M, "medium") to those of medium speed, Pi S (from S, "slow") to those of slow migration, Pi F (from F, "fast") to the rapid migration, and Pi Z, the last letter of the alphabet, to those with a very slow migration (Fig. 4.5).

The subsequent introduction of isoelectric focusing (IEF) techniques in ultrathin polyacrylamide gels and narrow pH gradients allowed to differentiate more accurately the AAT bands, and to identify approximately 30 new variants, which were designated with the letters A–L to those of faster and anodic migration and N–Z to those of slower and cathodic migration (compared with M) [15–17].

The most recent incorporation of sophisticated DNA amplification techniques and gene sequencing to characterize genotypes has risen to approximately 125 allelic variants, including several "rare" and "null" variants [18], which will be discussed in other chapters of the book.

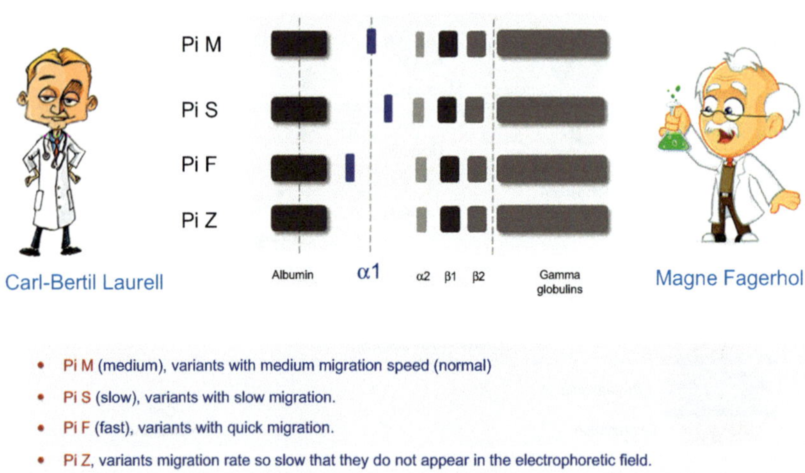

FIGURE 4.5 1967–68: the first classification of the Pi (protease inhibitor) system. After the discovery of alpha-1 antitrypsin, soon it was observed that the alpha-1 band presented many variants. Two pioneers in the history of the alpha-1 (Laurell and Faherhol) proposed grouping these variants on a classification system called "Pi" (protease inhibitor), on the basis of their migration velocity in starch gel electrophoresis (Pi M, medium; Pi S, slow; Pi F, fast; Pi Z, those variants with a migration rate so slow that they do not appeared in the electrophoretic field).

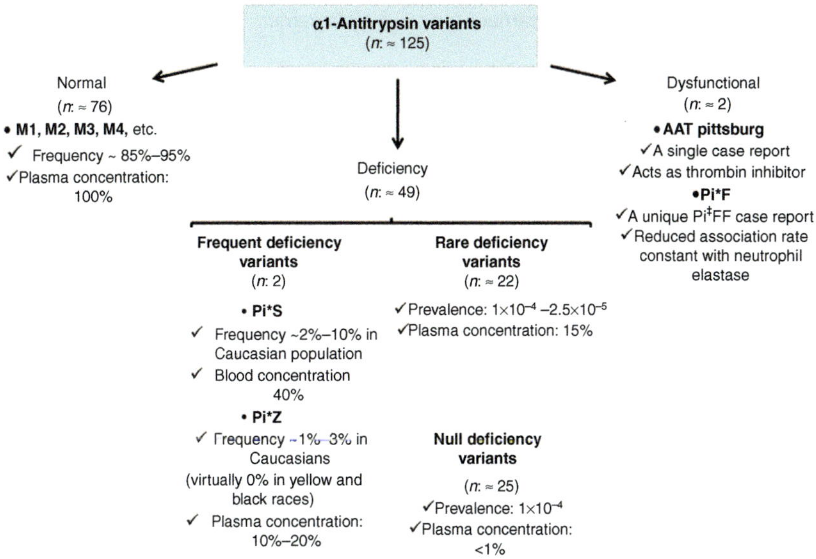

FIGURE 4.6 **The four major categories of the alpha-1 antitrypsin variants.** For clinical purposes, the AAT variants have been classified as normal, deficiency (subclassified into frequent, rare, and null), and dysfunctional variants.

Taking into consideration the large number of AAT variants, they have been classified into three major categories for clinical purposes (Fig. 4.6), namely:

1. Normal, characterized by normal AAT plasma concentration and not increased risk for AAT deficiency-related diseases. The most common of this group are M1, M2, M3, and M4 [1,19,20].
2. Deficiency, characterized by decreased AAT plasma concentration and increased risk for AAT deficiency-related diseases. Deficiency variants are also categorized as frequent (i.e., Pi*S and Pi*Z), rare (e.g., Mmalton), and null deficiency variants (e.g., Nullmadrid) [21].
3. Dysfunctional variants, including the AAT Pittsburg (a variant converted in antithrombin inhibitor) and the Pi*F variant, with a reduced association rate constant with neutrophil elastase [22].

4.3 GENETIC INHERITANCE

The AAT gene has two alleles, which are transmitted from parents to children by simple autosomal codominant Mendelian inheritance. Each child receives one allele from each parent. The normal allele is called M, and it is present in 85%–90% of individuals worldwide. Therefore, a normal individual receives two M alleles (one from each parent) forming a genotype MM [1].

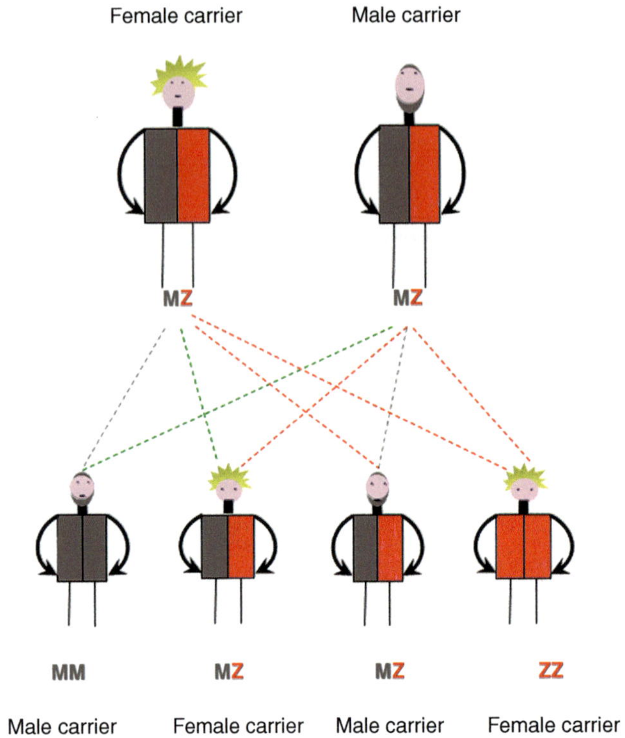

FIGURE 4.7 **Heritage: example of the alpha-1 antitrypsin gene transmission.** Both parents of this scheme are MZ, that is, they carry a normal allele (M) and a very deficient allelo (Z). In this case, their children have a 25% chance of receiving an MM genotype, a 50% of inheriting an MZ, and a 25% of receiving a ZZ genotype.

The most common deficient alleles are named S and Z, and they are, respectively, found in 2%–10% and 1%–3% of Caucasian populations of European ancestry. (A graphic example of the AAT gene transmission is shown in Fig. 4.7.)

As mentioned earlier, the vast majority of phenotypes found in practice are combinations of M, S, and Z alleles, namely MM (normal genotype exhibited by approximately 85%–90% of subjects), MS, SS, MZ, SZ, and ZZ (five deficiency genotypes found in the remaining 10%–15% of Caucasians). MM genotypes express 100% of AAT blood, whereas MS, SS, MZ, SZ, and ZZ express (roughly) 80, 60, 55, 40, and 15%, respectively [1,16,17].

Severe AAT deficiency defined by serum AAT levels below 35% of the average expected value, or 57–60 mg/dL (measured by nephelometry), or 11 μM (measured using the purified standard used in the US Registry), or 80 mg/dL (if the measurement is made by radial immunodiffusion, a disused technique), is a rare condition, usually associated with ZZ genotypes, and with less frequency with various combinations of Z, S, rare, and null alleles. In fact, in clinical practice 96% of the pathologies associated with AAT deficiency occur in homozygous ZZ, and the remaining 4% occur in heterozygous SZ, MZ, and the extremely infrequent "rare" and "null" genotypes [1] (Fig. 1.1).

4.4 NOMENCLATURE

The high number of variants and use of various methods for their characterization have motivated that to the classical acronym Pi the PI* symbol (with capital letters and asterisk) has been added to distinguish whether the determination has been made with IEF or PCR. Thus, when a phenotype identified (e.g., an MZ) at the protein level (i.e., by IEF) is called PiMZ, but if identified at the genetic level (with PCR techniques) the name would be PI*MZ. In this same example, PiMZ would be the phenotype, while PI*MZ would be the genotype. However, Sandhaus and Stoller on the 50th anniversary of the AAT deficiency discovery and the 25th anniversary of the Prolastina launch, with the intention of unifying this terminology, have proposed some basic terminology standards, such as using the AAT acronym to describe the alpha-1 antitrypsin; AATD to refer AAT deficiency; augmentative therapy rather than replacement therapy; Pi* (protease inhibitor, with "i" and asterisk) followed by the name of alleles (e.g., Pi*MZ; Pi*ZZ; Pi*Z/Null) to describe both phenotype and genotype; and Pi*Z when the diagnosis is made by IEF without sequencing the gene and therefore it is not possible to distinguish between a PiZZ and a PiZ/Null [23].

In recent years, the 28 alphabet letters have fallen short for classifying the increasing number of gene variants, and it has been necessary to use numerical suffixes to classify certain variants within a same phenotype (e.g., the M allele has been subclassified into M1, M2, M3, M4, and M5).

Null alleles are called simply "Null" or QO (a term of no clear meaning, adopted in Oslo in 1981 by an international committee of gene nomenclature, to designate genes not expressing proteins) [24], and both "null" and "rare" variants are named by a suffix with the name of the city or region where the first carrier was discovered (e.g., M_{malton}, S_{iiyama}, $Y_{barcelona}$, QO_{madrid}) [1,18,21].

Protease and peptidase terms are used interchangeably to designate the enzymes that break the protein peptide bonds. Depending on the substrate site of action, proteases are divided in turn into endopeptidases (or proteinases) that cleave peptide chains on the inside, and exopeptidases, which cleave the chains at terminal ends [25]. The proteinases are sorted according to the order of the catalytic residues of the peptide chain and flanking sequences into the following groups: serine-proteinases, threonine-proteinases, cysteine-proteinases, aspartyl-proteinases, metallo-proteinases, glutamyl-proteinases, and "mixed" [26,27]. According to this scheme, elastase (the specific substrate of AAT) would be a protease or peptidase, in general terms; more specifically, a proteinase or endopeptidase, and specifically a serine proteinase.

The nomenclature to describe subsites in peptidases was introduced in 1967 by Schechter and Berger [28]. According to this system, the amino acid residues

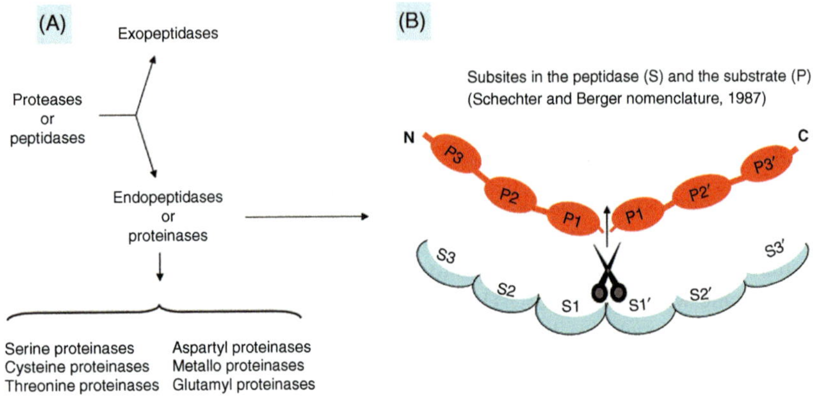

FIGURE 4.8 Classification (A) and site of action of proteases in the peptide chain of the substrate (B). (A) Nomenclature of the International Union of Biochemistry (IUB), 1964; (B) nomenclature of Schechter and Berger.

of the substrate bind to the subsites of the enzyme active site. Subsites in the peptidases are called S (from "subsite") and the amino acid residues in the substrate P (from "peptide"). The link to be cut is between P1 and P1′. Residues toward the N-terminal end of the substrate are numbered as P1, P2, P3, and so on, and to the C-terminal end as P1′, P2′, P3′, while the complementary protease subsites are similarly numbered as S1, S2, S3 and S1′, S2′, S3′, and so on (Fig. 4.8).

REFERENCES

[1] American Thoracic Society/European Respiratory Society Statement. Standards for the diagnosis and management of individuals with alpha1-antitrypsin deficiency. Am J Respir Crit Care Med 2003;168:818–900.

[2] Long GL, Chandra T, Woo SLC, Davie EW, Kurachi K. Complete sequence of the cDNA for human alpha-1-antitrypsin and the gene for the S variant. Biochemistry 1984;23:4828–37.

[3] Perlino E, Cortese R, Ciliberto G. The human alpha-1-antitrypsin gene is transcribed from two different promoters in macrophages and hepatocytes. EMBO J 1987;6:2767–71.

[4] Hafeez W, Ciliberto G, Perlmutter DH. Constitutive and modulated expression of the human alpha 1 antitrypsin gene. Different transcriptional initiation sites used in three different cell types. J Clin Invest 1992;89:1214–22.

[5] Rollini P, Fournier RE. Differential regulation of gene activity and chromatin structure within the human serpin gene cluster at 14q32.1 in macrophage microcell hybrids. Nucleic Acids Res 2000;28:1767–77.

[6] Kalsheker NA, Morgan K. Regulation of the alpha 1-antitrypsin gene and a disease-associated mutation in a related enhancer sequence. Am J Respir Crit Care Med 1994;150:S183–9.

[7] Kalsheker N, Morley S, Morgan K. Gene regulation of the serine proteinase inhibitors alpha1-antitrypsin and alpha1-antichymotrypsin. Biochem Soc Trans 2002;30:93–8.

[8] Yamamoto K, Sinohara H. Regulation by sex hormones of serum levels of contrapsin and alpha 1-antiprotease in the mouse. Biochim Biophys Acta 1984;798:231–4.

[9] Perlmutter DH, Travis J, Punsal PI. Elastase regulates the synthesis of its inhibitor, alpha 1-proteinase inhibitor, and exaggerates the defect in homozygous PiZZ alpha 1 PI deficiency. J Clin Invest 1988;81:1774–80.

[10] Knoell DL, Ralston D, Coulter KR, Wewers MD. Alpha 1-antitrypsin and protease complexation is induced by lipopolysaccharide, interleukin-1beta, and tumor necrosis factor-alpha in monocytes. Am J Respir Crit Care Med 1998;157:246–55.

[11] Lisowska-Myjak B, Pachecka J. Antigenic and functional levels of alpha-1-antitrypsin in serum during normal and diabetic pregnancy. Eur J Obstet Gynecol Reprod Biol 2003;106:31–5.

[12] Brueckner F, Armache KJ, Cheung A, Damsma GE, Kettenberger H, Lehmann E, et al. Structure-function studies of the RNA polymerase II elongation complex. Acta Crystallogr D Biol Crystallogr 2009;65:112–20.

[13] Köhler A, Hurt E. Exporting RNA from the nucleus to the cytoplasm. Nat Rev Mol Cell Biol 2007;8:761–73.

[14] Fagerhol MK. The Pi system: genetic variants of serum alpha-1-antitrypsin. Ser Haematol 1968;1:153–61.

[15] Cox DW. Genetic variation in alpha-1-antitrypsin. Am J Hum Genet 1978;30:660–2.

[16] Donato LJ, Jenkins SM, Smith C, Katzmann JA, Snyder MR. Reference and interpretive ranges for α(1)-antitrypsin quantitation by phenotype in adult and pediatric populations. Am J Clin Pathol 2012;138:398–405.

[17] Bornhorst JA, Greene DN, Ashwood ER, Grenache DG. α1-Antitrypsin phenotypes and associated serum protein concentrations in a large clinical population. Chest 2013;143:1000–8.

[18] Lee JH, Brantly M. Molecular mechanisms of alpha1-antitrypsin null alleles. Respir Med 2000;94(Suppl. C):S7–S11.

[19] Blanco I, Fernández-Bustillo E, de Serres FJ, Alkassam D, Rodríguez Menéndez C. PI*S and PI*Z alpha 1-antitrypsin deficiency: estimated prevalence and number of deficient subjects in Spain. Med Clin (Barc) 2004;123:761–5.

[20] Piras B, Ferrarotti I, Lara B, Martinez MT, Bustamante A, Ottaviani S, et al. Clinical phenotypes of Italian and Spanish patients with α1-antitrypsin deficiency. Eur Respir J 2013;42:54–64.

[21] Ferrarotti I, Thun GA, Zorzetto M, Ottaviani S, Imboden M, Schindler C, et al. Serum levels and genotype distribution of α1-antitrypsin in the general population. Thorax 2012;67:669–74.

[22] Lara B, Martínez MT, Blanco I, Hernández-Moro C, Velasco EA, Ferrarotti I, et al. Severe alpha-1 antitrypsin deficiency in composite heterozygotes inheriting a new splicing mutation QOMadrid. Respir Res 2014;15:125.

[23] Sinden NJ, Koura F, Stockley RA. The significance of the F variant of alpha-1-antitrypsin and unique case report of a PiFF homozygote. BMC Pulm Med 2014;14:132.

[24] Sandhaus RA, Stoller JK. Introduction to the 50th anniversary of the description of alpha-1 antitrypsin deficiency. COPD 2013;10(Suppl. 1):1–2.

[25] Shows TB, McAlpine PJ. The 1981 catalogue of assigned human genetic markers and report of the nomenclature committee. Oslo Conference (1981): Sixth International Workshop on Human Gene Mapping. Cytogenet Cell Genet 1982;32:221–45.

[26] Quesada V, Ordóñez GR, Sánchez LM, Puente XS, López-Otín C. The Degradome database: mammalian proteases and diseases of proteolysis. Nucleic Acids Res 2009;(Database issue):D239–43.

[27] López-Otín C, Bond JS. Proteases: multifunctional enzymes in life and disease. J Biol Chem 2008;283:30433–7.

[28] Schechter I, Berger A. On the size of the active site in proteases. I. Papain. Biochem Biophys Res Commun 1967;27:157–62.

Chapter 5

Alpha-1 Antitrypsin Deficiency: Liver Pathophysiology

5.1 INTRODUCTION

The pathophysiology of liver injury in alpha-1 antitrypsin deficiency (AATD) is complex and only partially known. Nevertheless, it is practically accepted by all experts on this field that the abnormal accumulation of alpha-1 antitrypsin (AAT) polymers in the rough endoplasmicum reticulum (RER) of hepatocytes incites a cellular stress response, which finally can cause liver damage [1].

In clinical practice, the vast majority of cases with liver disease related to AATD are homozygous ZZ, only a small number are heterozygous SZ, and anecdotally, there have been reported isolated cases of liver disease in Rare/Z alleles compound heterozygous, such as I/Z [2,3]. There have also been described several rare deficient alleles [4–36], some of which as Siiyama, Mmalton, Mduarte, and so on secrete abnormal proteins, resembling Z-protein, that polymerize into hepatocytes, being theoretically able to cause liver damage in Rare/Rare homozygotes or Z/Rare compound heterozygotes. However, the scarcity of case reports and the lack of monitoring data make it impossible an adequate knowledge of the clinical behavior and natural history on these rare alleles (Table 5.1).

5.2 THE Z AND S MUTATIONS

The complicated structure of the AAT molecule not only gives it significant inhibitory advantages, but also makes it very vulnerable to conformational changes associated with various deficiency mutations, the most frequent S and Z [37].

The exon sequence of the mutated Z gene differs only of the normal M gene in a nitrogenous base of a codon in exon V, consisting of GAG (guanine–adenine–guanine) instead of AAG (adenine–adenine–guanine). The replacement of guanine by adenine in the triplet changes the protein amino acid sequence by transcribing a lysine molecule rather than a normal glutamic acid residue at position 342 of the polypeptide chain (i.e., GAG Glu342 $\rightarrow$ AAG Lys342) [1,2].

In the S mutation, there is a change of a base of the normal GAA triplet (guanine–adenosine–adenosine) by GTA (guanine–thymine–adenosine) at exon III. Replacement of adenosine by thymine in this codon causes transcription of

TABLE 5.1 List of Alpha-1 Antitrypsin Deficiency "Rare" Alleles (in Alphabetical Order)

Allele	Mutation and pathophysiology	Clinical details and potential risk	Author, country, and year [reference]
Etaurisano	Lys259Ile. Intrahepatic polymerization with secondary reduction of plasma AAT concentration	One index case with emphysema. High risk of emphysema and liver disease in homozygotes and compound heterozygotes	Fra, Italy, 2012 [4]
F	TGT223GGT. Normal serum levels of AAT, but very low association constant with elastase and proteinase 3 (low inhibitory capacity)	A PiFF with COPD, various heterozygous FZ with lung emphysema, and 1 heterozygous FZ with panniculitis have been reported	Sinden, UK, 2014 [5]; Suh-Lailam, EEUU, 2014 [6]; Elsensohn, EEUU, 2015 [7]
I	Exon II. Arg39Cis. I protein forms heteropolymers with Z into hepatocytes, secondarily expressing reduced AAT serum levels	Isolated cases, including a patient with liver I/Z heteropolymers and cirrhosis. It is believed that homozygous are not at risk, but they do compound heterozygotes	Graham, UK, 1989 [8]; Seri, Italy, 1992 [9]; Mahadeva, UK, 1999 [3]
King's	His334Asn. Intrahepatic polymerization and reduced AAT serum level similar to Z	A single case with neonatal jaundice. Probable increased risk for emphysema and cirrhosis in homozygotes and compound heterozygotes	Miranda, UK, 2010 [10]
Mbaghdad	Ala336Pro. It affects a residue of the β band 5-A of the "shutter" which causes polymerization, reduced serum (predominantly polymeric AAT) and severe loss of antiprotease capacity	One homozygous case detected in an asymptomatic woman of Iraqi origin, with AAT levels of 0.4 g/L. Homozygotes with this mutation may have a theoretical risk comparable to that of ZZ	Haq I, UK, 2015 [11]

TABLE 5.1 **List of Alpha-1 Antitrypsin Deficiency "Rare" Alleles (in Alphabetical Order)** ***(cont.)***

Allele	Mutation and pathophysiology	Clinical details and potential risk	Author, country, and year [reference]
Mheerlen	Exon V. Pro369Leu. Intrahepatic polymerization causing low serum AAT concentration	Isolated cases. High risk of emphysema and liver disease in homozygotes and compound heterozygotes	Hofker, Holland, 1989 [12]; Poller, Germany, 1999 [13]
Mmalton	Exon II. 52Phe 3 bp deletion. Intrahepatic degradation and polymerization and reduced serum levels	Isolated cases in Europe, EEUU and Canada. High risk (similar to Pi*Z) of emphysema and liver disease in homozygotes and compound heterozygotes	Fraizer, Canada, 1989 [14]; Curiel, EEUU, 1989 [15]; Cox, Canada, 1989 [16]; Graham, UK, 1989 [8] Denden, Tunisia, 2009 [17]; Lara, Spain, 2013 [18]; Loly, France, 2015 [19]
Mmineral springs	Exon II. Gly67Glu. Abnormal protein synthesis without polymerization. Low serum levels and markedly reduced ability to inhibit elastase	One case detected in the United States in an Afro-Caribbean subject with emphysema. High risk of emphysema in homozygous and compound heterozygous	Curiel, EEUU, 1990 [20]
Mnichinan	Exon II. Deletion 52Phe and Gly148Arg. Intrahepatic polymerization, and very reduced serum concentration	Described in a Japanese family with blood relationship. High risk of liver disease and emphysema.	Matsunaga, Japan, 1990 [21]
Mpalermo	Exon II. Deletion 51Phe. Unknown intracellular defect. Severe plasmatic deficiency	A single case with emphysema. High risk of emphysema in homozygous and compound heterozygous	Faber, Germany, 1994 [22]
Mpisa	Lys259Ile. Intrahepatic polymerization and reduced serum levels of AAT	A single case. High risk of emphysema and liver disease in homozygotes and compound heterozygotes	Fra, Italy, 2012 [4]

(Continued)

TABLE 5.1 List of Alpha-1 Antitrypsin Deficiency "Rare" Alleles (in Alphabetical Order) (*cont.*)

Allele	Mutation and pathophysiology	Clinical details and potential risk	Author, country, and year [reference]
Mprocida	Exon II. Leu41Pro. Unstable protein mostly degraded in liver without being secreted. Elastase activity greatly diminished	A single case reported. High risk of emphysema and liver disease in homozygotes and compound heterozygotes	Takahashi, EEUU, 1988 [23]
Mvall d'hebron (=wurzburg)	Pro369Ser. Protein retention in liver by undefined mechanisms. Reduced serum levels	Isolated cases in Spain, Germany, Switzerland, and Tunisia. High risk of emphysema in homozygous and compound heterozygous	Jardi, Spain, 2000 [24]; Poller, Germany, 1999 [13]; Zorzetto, Switzerland, 2008 [25]; Denden, Tunisia, 2009 [17]
Mvarallo	Deletion 41–51 and stop codon 70–71. Unknown intracellular defect. Reduced serum levels	A heterozygous M/Mvarallo with 50% serum AAT. High risk of emphysema in homozygous and compound heterozygous	Coni, Italy, 2003 [26]
Pittsburgh	Met358 Arg. Dysfunctional AAT with antithrombin-3 activity	A single case in the World of a 14-year-old subject who died from fatal bleeding diathesis	Owen, New Zealand, 1983 [27]
Pduarte	Exon III. Asp256Val (M4). Intracellular degradation with secondary plasmatic AAT deficiency	Isolated cases. High risk of emphysema in homozygous and compound heterozygous	Graham, UK, 1989 [8]; Faber, Germany, 1989 [28]; Hildesheim, EEUU, 1993 [29]
Plowell (=QOcardiff)	Exon III. Asp256Val (M3). Mutation similar to Duarte in M3. Intracellular degradation and reduced serum levels of AAT	Isolated cases. High risk of emphysema in homozygous and compound heterozygous	Holmes, EEUU, 1970 [30]; Seri, Italy, 1992 [9]

TABLE 5.1 List of Alpha-1 Antitrypsin Deficiency "Rare" Alleles (in Alphabetical Order) (*cont.*)

Allele	Mutation and pathophysiology	Clinical details and potential risk	Author, country, and year [reference]
Siiyama	Exon II. Ser53Phe. Intrahepatic polymerization similar to Z protein, causing very low AAT serum concentration	Several index cases in 13 Japanese families with frequent inbreeding. High risk of liver disease and emphysema in homozygous and compound heterozygous	Yuasa, Japan, 1993 [31]; Seyama, Japan, 1995 [32]
Wbethesda	Exon V. Ala336Thr. Intracellular degradation. Low serum AAT level	An isolated case. High risk of emphysema in homozygous and compound heterozygous	Holmes, EEUU, 1990 [33]
Ybarcelona	Asp256Val y Pro391His. Unknown intracellular defect. Very reduced serum AAT concentration	Reported a homozygote with severe emphysema and several asymptomatic relatives. High risk of emphysema in homozygous and compound heterozygous	Jardi, Spain, 1998 [34]
Yorzinuovi	Pro391His. Intrahepatic polymerization and reduced AAT serum levels	One case reported. High risk of emphysema and liver disease in homozygotes and compound heterozygotes	Fra, Italy, 2012 [4]
Zausburg (=Ztun)	Exon V. Glu342 Lys (M2). Intrahepatic degradation and polymerization. Reduced serum AAT concentration	An isolated case. High risk of liver disease and emphysema in homozygotes and compound heterozygotes	Weidinger, Germany, 1985 [35]; Whitehouse, UK, 1989 [36]
Zwrexham	Ser19 Leu y Glu342Lys (Z mutation), in M2. Unknown intracellular defect. Decreased serum levels (as a Pi*Z)	One heterozygote Z/Zwrexham with emphysema. High risk for emphysema in homozygotes and compound heterozygotes	Graham, UK, 1989 [8]

Ala, Alanine; Arg, arginine; Asn, asparagine; Asp, aspartic acid; Bp, base pairs; Cys, cisteine; EEUU, the United States; Glu, gutamic acid; Gly, glycine; His, histidine; Ile, isoleucine; Leu, leucine; Lys, lysine; Met, methionine; Phe, phenylalanine; Pro, proline; Ser, serine; Thr, trheonine; Val, valine.

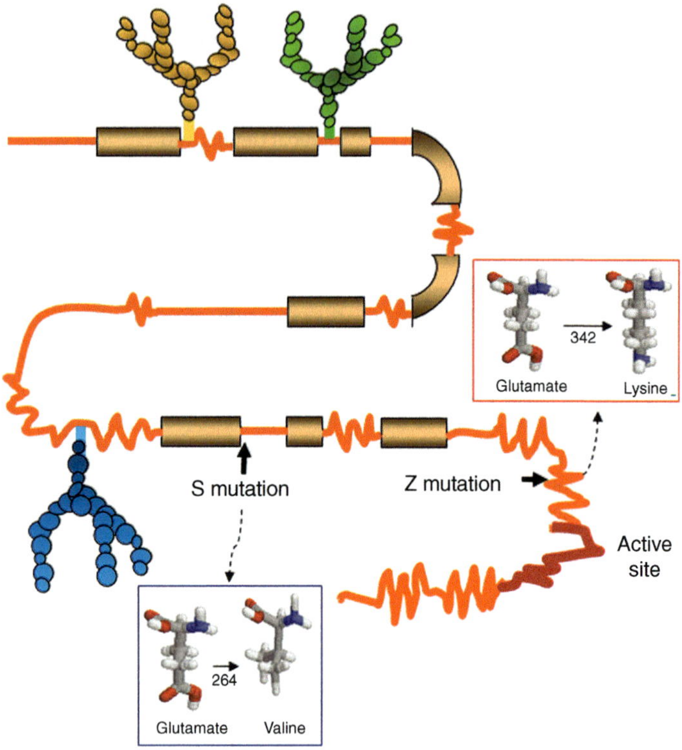

FIGURE 5.1 Location of S and Z mutations in the alpha-1-antitrypsin molecule. In this schematic representation, with the protein unrolled it can be see that in the S mutation there is a replacement of glutamic by valine at position 264 far away from the active site, a change that significantly preserves the integrity of the molecule. However, in the Z mutation the change of a single amino acid (glutamic acid for lysine) in the insertion point of one end of the active site loop in the upper region of an A sheet causes pronounced structural changes having significant pathological consequences.

a valine molecule instead of glutamic acid at position 264 (i.e., GAA Glu264 → GTA Lys264) (Fig. 5.1).

In the Z mutation, the change of a single amino acid (glutamic acid for lysine) in the insertion point of one end of the active site loop in the upper region of an A sheet causes pronounced structural changes with significant pathological consequences [38].

In the S mutation, the single amino acid replacement occurs away from the active site loop, causing less pronounced conformational changes in the protein structure [39].

Z mutation generates unstable proteins, which have a conformation called metastable (M*), which is characterized by abnormal widening of the physiological surface cavities "breach" and "shutter" of the molecule, the fall of the active site loop over A sheets of the protein, and the advancement of a helix to

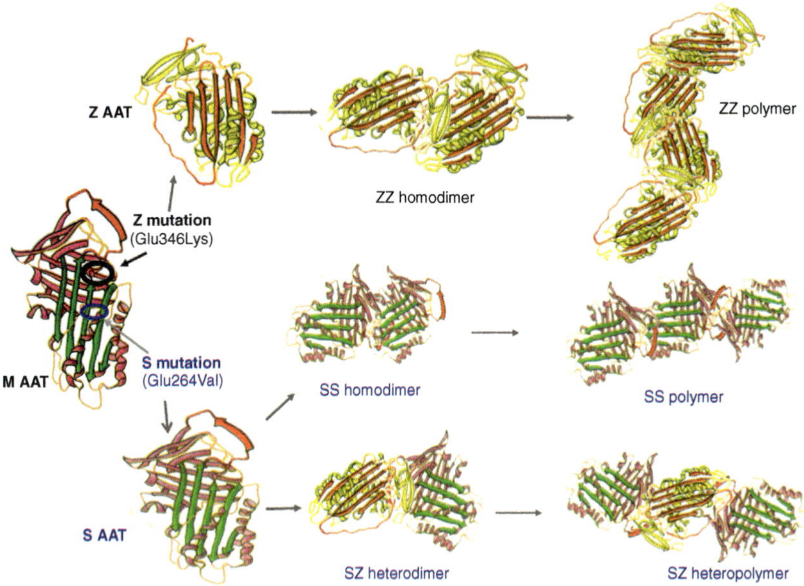

FIGURE 5.2 Z and S mutations, conformational changes, and polymerization. The Z mutation generates unstable proteins, which have a conformation called metastable (M*), characterized by an abnormal widening of the physiological cavities of its surface, the fall of the active site loop over the A sheets, and the advancement of a helix to close the gap created at the bottom of the molecule. From the metastable state M*, the Z protein readily passes to a delta (δ) latent state, and from this back to M*, which tends to aggregate with other mutated molecules by inserting its active site in the enlarged cavities of the A sheets of other to form dimers, which in turn tend to bind with other mutated molecules to form large polymers arranged in one-by-one rows, like beads on a string. In the S mutation, the active site loop remains exposed, and although the S proteins can polymerize, this polymerization is slower and less firm than the Z one, and is easily controlled by the potent clearance mechanisms of the cell. However, SZ heterozygous can form compact, durable, and toxic heteropolymers.

close the gap created between A sheets at the bottom of the molecule. From the conformational M* state, Z protein readily passes to a delta (δ) latent state, and from this backs to M*, which tends to aggregate with other mutated molecules by inserting its active site in the enlarged cavities between A sheets of other molecules to form dimers, which in turn tend to bind with other Z molecules to form large polymers arranged in one-by-one rows, like beads on a string [38] (Fig. 5.2).

5.3 PATHOPHYSIOLOGY

The precise mechanisms underlying the formation of polymers are not known with accuracy, although they are the subject of an active research to find solutions for various diseases characterized by abnormal accumulation of misfolded proteins in different tissues, such as amyloidosis, Alzheimer's, prion encephalopathy, serpinopathies, and other.

During the multistep, enzyme-catalyzed process of biosynthesis, the Z protein is correctly transcribed and translated by ribosomes, and the nascent Z polypeptide chain translocated into the hepatocyte RER lumen. Then, the "quality control system" of the RER, basically consisting of mannosidase 1, the chaperones calnexin and ubiquitin, and possibly other not well-identified proteins, residents in the cell RER, bind sequentially to the nascent polypeptide chains to cooperate in their correct folding (Fig. 5.3). However, most mutated chains of Z AAT (80%–90%) cannot be properly folded, forming dimers and polymers among them, which are retained by the "quality control system." Mannosidase 1 easily detects the mutated molecules, and chaperones join them to form complexes that are cleared by various intracellular proteolytic systems that successfully process the vast majority of Z molecules retained in the RER.

The best-known proteolytic systems are the ubiquitin-dependent proteasome, the ubiquitin-independent proteasome, and the autophagy pathway, but possibly there are other degradation systems still unidentified (Fig. 5.4).

In the ubiquitin-dependent pathway, proteins are tagged for degradation with ubiquitin by a reaction catalyzed by ubiquitin ligases. The end result is a polyubiquitin chain that is bound by the proteasome, allowing it to degrade the tagged proteins after they are deubiquitinated. Trimolecular calnexin–AAT–ubiquitin complexes are also formed to be finally degraded by proteasome proteases. The degradation process produces short peptides which can then be further degraded into amino acid sequences and reused by cells in synthesizing new proteins.

Although most proteasomal substrates must be initially ubiquitinated and then deubiquitinated just before being degraded, some of misfolded, or highly oxidized, proteins can also be degraded by a proteasome ubiquitin-independent pathway, especially under intense cellular stress conditions.

Furthermore, large accumulations of polymerized proteins are preferably removed by autophagy mechanisms linked to lysosomes rather than by proteasome. This mechanism involves the formation of a double-membrane vesicle about the polymers marked for destruction, known as autophagosome. Autophagosome joins and merges with lysosomes in the cytoplasm, and so the autophagosome contents are degraded by lysosomal hydrolases. The final degradation products are recycled for reuse (Fig. 5.4).

Liver polymerization is a capricious and erratic phenomenon that occurs with great intensity in some hepatocytes, while its intensity is low or does not occur at all in most of the other hepatic cells. Consequently, the accumulation of Z protein in the liver is not homogeneous, and while some hepatocytes can show large accumulations of proteins in the same liver, accumulations are scarce or do not occur. The cause of this intriguing differences is unknown, although it

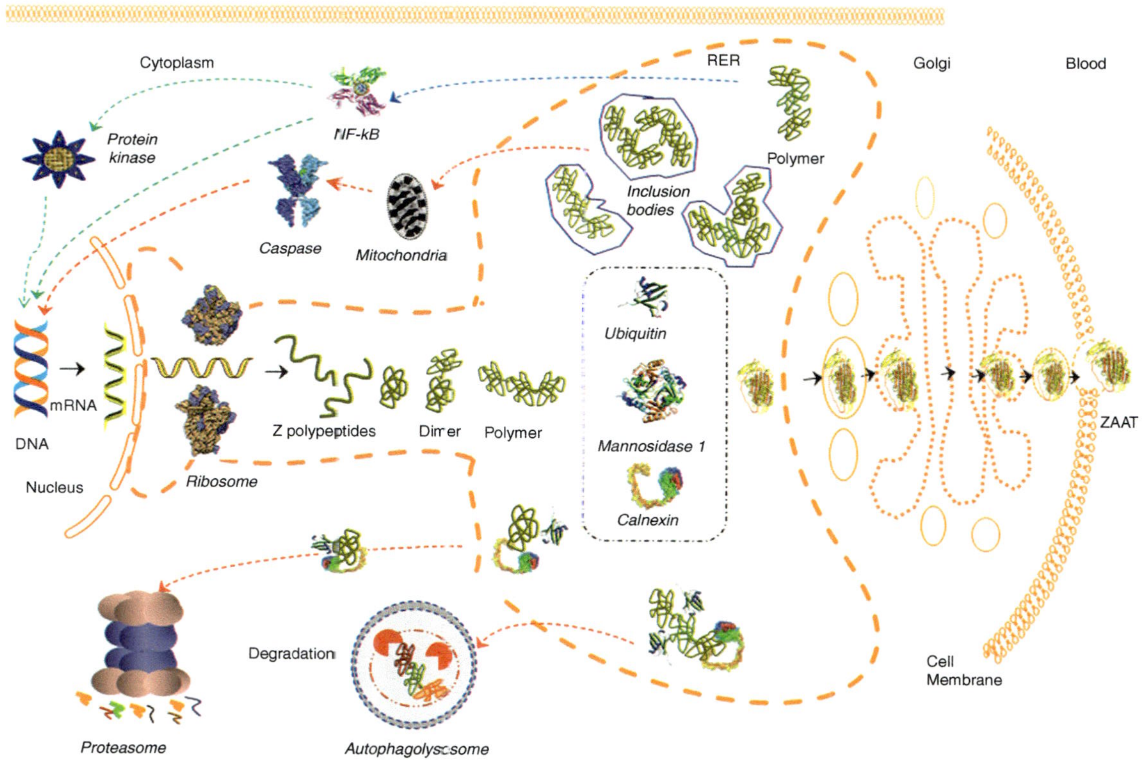

FIGURE 5.3 **Intracellular pathophysiological events related to the mutant alpha-1 antitrypsin Z.** Abnormal accumulation of alpha-1 antitrypsin (AAT) polymers in the rough endoplasmic reticulum (RER) of hepatocytes incites a cellular stress response, which finally can cause liver damage. Polypeptides and polymers are detected by mannosidase 1 and marked by calnexin and ubiquitin to be degraded via proteasome or autophagia. Only a small number of molecules partially folded through the Golgi apparatus and are secreted into the blood. Polymer-inclusion bodies activate various elements of the cytoplasm and nucleus, accelerating apoptosis and cell cycle rate, and eventually leading to liver fibrosis, cirrhosis, or hepatocarcinoma.

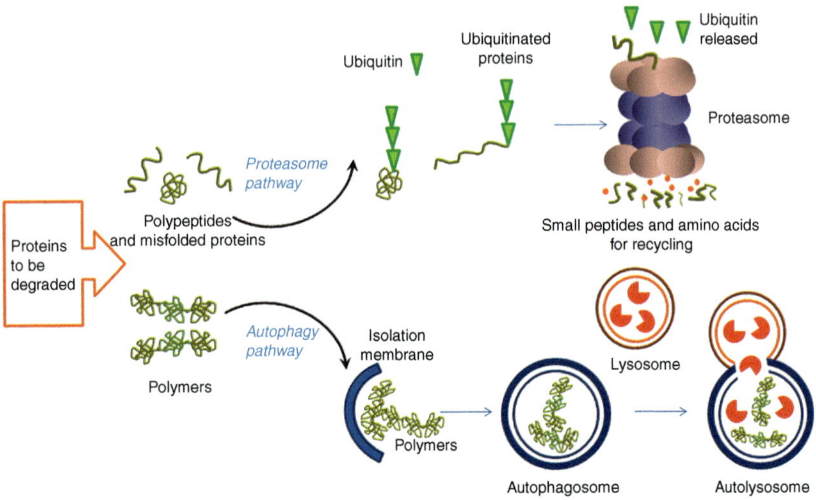

FIGURE 5.4 Degradation pathways of Z proteins. In the hepatocyte as in any other eukaryotic cell, two major pathways, the ubiquitin-proteasome pathway and the lysosomal proteolysis (autophagia), mediate protein degradation. In the ubiquitin-proteasome pathway, proteins are marked for rapid degradation by several molecules of ubiquitin. Then, misfolded proteins are unfolded and targeted to proteasome, where ubiquitin is released and the peptidic chains are broken down into smaller components. Autophagy involves the formation of vesicles (autophagosomes) where polymers are enclosed in membranes. Autophagosomes fuse with lysosomes, and then lysosomal enzymes digest the polymers.

is assumed that may be linked to the different capacity of AAT secretion by the population of hepatocytes of a same liver.

In hepatocytes in which the polymerization is intense, polymers may form very stable inclusion bodies surrounded by membranes. These inclusion bodies emit depolarization signals to mitochondrias, and these release cytochrome C, which in turn activates the caspase cascade, and this simultaneously stimulates cell apoptosis and autophagy. Continued cellular stress causes mitochondrial damage following these sequential processes: (1) ATP depletion, (2) calcium homeostasis alteration, (3) oxidative stress, and (4) increased permeability of mitochondrial membranes with cytochrome C release. On the contrary, Z-polymers emit signals for the cytosolic nuclear factor kappa B transcription factor, which leads to a compensatory activation of protein kinases to accelerate the rate of cell cycle in an attempt to counteract the apoptotic activity of caspases. The nuclear factor kappa B can also cross the nuclear membrane and act directly on the DNA as a transcription factor that accelerates the cell cycle [39].

Eventually, the uninterrupted successive cycles of liver damage with cell death in hepatocytes containing high loads of polymers and inclusion bodies, and the compensatory proliferation of other hepatocytes less loaded of polymers,

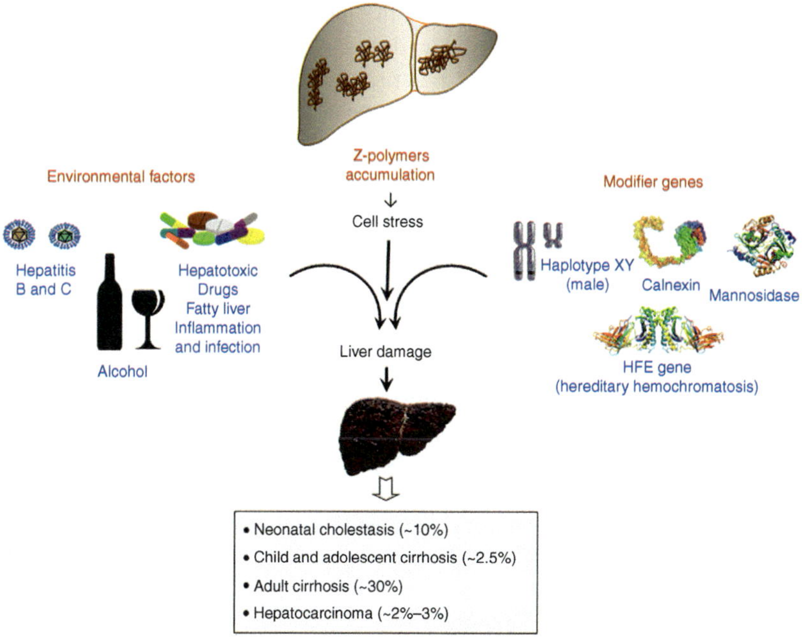

FIGURE 5.5 Alpha-1 antitrypsin deficiency: a complex monogenic disorder with many types of clinical manifestations. Liver accumulation of polymers can cause liver damage, which can manifest clinically by various types of liver diseases in children, youth, and adults. Given the large interindividual variability of the clinical expression of the liver disease, it is thought that it can significantly influence several exogenous factors and/or other genetic modifiers not well identified so far.

can generate an imbalance between apoptosis and regeneration, which can lead to necrosis, fibrosis, or cirrhosis of the damaged organ, and even to the development of tumors (Fig. 5.5).

Although in the Z mutation intracellular polymerization occurs in 85%–90% of the molecules in the S mutation, polymerization occurs at most in 60% of them. In addition, protein S polymerizes more slowly than Z, and can be cleared more easily by the cell preventing it to form inclusion bodies, which are the most important triggers of cellular stress and liver damage. Indeed, homozygous SS do not develop liver disease because they do not form inclusion bodies within hepatocytes. However, heteropolymers made of protein S and Z have been found in heterozygous SZ, and in fact some SZ heterozygotes may occasionally develop liver cirrhosis identical to that of the ZZ homozygotes. I protein can also form heteropolymers with protein Z in the hepatocyte in isolated cases, as it has been reported in a male aged 34 with liver cirrhosis associated with I/Z heteropolymers [3]. In this same context, hepatocytes of MZ carriers do not show significant accumulation of polymers and do not develop liver disease, suggesting that the implementation of the mechanisms for cell injury is directly dependent on the load of protein deposits. Theoretically, mutant proteins produced by some rare alleles, as Etaurisano, Mheerlen, I, King's, Mmalton, Mpisa, Siiyama, Yorzinuovi, Zausburg, Ztun, or Zwrexham, can polymerize in liver, in a similar way as Z, but for its extreme rarity there is no evidence of this happening (Table 5.1).

Molecules that overcome the RER quality control (approximately 40% of S and 15% of Z) pass to the Golgi apparatus, where they are packaged into secretory vesicles to be glycosylated and purified before being discharged into the bloodstream [39]. However, in the case of the Z protein the few molecules that reach the bloodstream have metastable, polymeric, and latent conformations, with an overall 80% loss of their inhibitory capacity [40]. By contrast, as the active site of S protein remains unaffected, it retains its inhibitory capacity unchanged [41]. Remarkably, although its meaning is still unknown, a recent study on circulating AAT polymers of hepatic origin has found that they represented a 15% of all the circulating AAT in Pi*ZZ subjects, but much minor percentage in MZ and SZ heterozygotes, and a negligible in MM, MS, and SS genotypes [42].

5.4 CLINICAL EXPRESSION

In general, AATD is a congenital condition with marked interindividual variability in clinical presentation. Thus, although a high percentage of Pi*ZZ subjects may develop pulmonary emphysema, other can develop liver cirrhosis; very few, both diseases at once; a minority other less typical pathologies, such as panniculitis or vasculitis, while more than a third of them shows no clinical symptoms throughout their lifetime [37].

The reason for this variability is not known, but it seems clear that AATD is not a disease itself but a complex monogenic disorder that predisposes to the development of various diseases. For this to happen, it must generally concur in the same individual, in addition to the congenital protein deficiency, other favoring external factors, and/or other not well-known modifier genes [43]. For example, it is thought that alcohol and other hepatotoxic substances, fatty liver, viral hepatitis B and C, recurrent inflammatory processes, or infections (which increase Z AAT synthesis and polymers formation) can be favoring external factors of liver damage.

On the contrary, no evidence is available to explain why only some individuals with AATD and apparently in a same environmental conditions develop liver diseases, while most do not. It is suspected that (apart from the probably favoring factors associated with male sex) there can be associated other unknown genetic mutations. Among these possible favoring candidate genes, they have been cited the calnexin gene, the mannosidase 1 gene, and the human hemochromatosis gene [43–45] (Fig. 5.5).

REFERENCES

[1] Lomas DA, Mahadeva R. Alpha1-antitrypsin polymerization and the serpinopathies: pathobiology and prospects for therapy. J Clin Invest 2002;110:1585–90.

[2] Salahuddin P. Genetic variants of alpha1-antitrypsin. Curr Protein Pept Sci 2010;11:101–17.

[3] Mahadeva R, Chang WS, Dafforn TR, Oakley DJ, Foreman RC, Calvin J, et al. Heteropolymerization of S, I, and Z α1-antitrypsin and liver cirrhosis. J Clin Invest 1999;103:999–1006.

[4] Fra AM, Gooptu B, Ferrarotti I, Miranda E, Scabini R, Ronzoni R, et al. Three new alpha1-antitrypsin deficiency variants help to define a C-terminal region regulating conformational change and polymerization. PLoS One 2012;7:e38405.

[5] Sinden NJ, Koura F, Stockley RA. The significance of the F variant of alpha-1-antitrypsin and unique case report of a PiFF homozygote. BMC Pulm Med 2014;14:132.

[6] Suh-Lailam BB, Procter M, Krautscheid P, Haas J, Kumar S, Mao R, et al. Challenging identification of a novel PiISF and the rare PiMmaltonZ α1-antitrypsin deficiency variants in two patients. Am J Clin Pathol 2014;141:742–6.

[7] Elsensohn AN, Liaqat M, Powell G, Duffy KL, Bowen AR, Petersen MJ. Alpha-1-antitrypsin associated panniculitis and corneal ulcer with a rare allelic variant (PiFZ): a case report. Int J Dermatol 2016;55:698–9.

[8] Graham A, Kalsheker NA, Newton CR, Bamforth FJ, Powell SJ, Markham AF. Molecular characterisation of three alpha-1-antitrypsin deficiency variants: proteinase inhibitor (Pi) nullcardiff (Asp256—Val); PiMmalton (Phe51—deletion) and PiI (Arg39—Cys). Hum Genet 1989;84:55–8.

[9] Seri M, Magi B, Cellesi C, Olia PM, Renieri A, De Marchi M. Molecular characterization of the P and I variants of alpha 1-antitrypsin. Int J Clin Lab Res 1992;22:119–21.

[10] Miranda E, Pérez J, Ekeowa UI, Hadzic N, Kalsheker N, Gooptu B, et al. A novel monoclonal antibody to characterize pathogenic polymers in liver disease associated with α1-antitrypsin deficiency. Hepatology 2010;52:1078–88.

[11] Haq I, Irving JA, Saleh AD, Dron L, Regan-Mochrie GL, Motamedi-Shad N, et al. Deficiency mutations of alpha-1 antitrypsin. Effects on folding, function, and polymerization. Am J Respir Cell Mol Biol 2016;54:71–80.

[12] Hofker MH, Nukiwa T, van Paassen HM, Nelen M, Kramps JA, Klasen EC, et al. A Pro—Leu substitution in codon 369 of the alpha-1-antitrypsin deficiency variant PI MHeerlen. Hum Genet 1989;81:264–8.

[13] Poller W, Merklein F, Schneider-Rasp S, Haack A, Fechner H, Wang H, et al. Molecular characterization of the defective alpha 1-antitrypsin alleles PI Mwurzburg (Pro369Ser), Mheerlen (Pro369 Leu), and Q0lisbon (Thr68Ile). Eur J Hum Genet 1999;7:321–31.

[14] Fraizer GC, Harrold TR, Hofker MH, Cox DW. In-frame single codon deletion in the Mmalton deficiency allele of a1-antitrypsin. Am J Hum Genet 1989;44:894–902.

[15] Curiel DT, Holmes MD, Okayama H, Brantly ML, Vogelmeier C, Travis WD, et al. Molecular basis of the liver and lung disease associated with the alpha 1-antitrypsin deficiency allele Mmalton. J Biol Chem 1989;264:13938–45.

[16] Cox DW, Billingsley GD. Rare deficiency types of alpha 1-antitrypsin: electrophoretic variation and DNA haplotypes. Am J Hum Genet 1989;44:844–54.

[17] Denden S, Zorzetto M, Amri F, Knani J, Ottaviani S, Scabini R, et al. Screening for alpha 1 antitrypsin deficiency in Tunisian subjects with obstructive lung disease: a feasibility report. Orphanet J Rare Dis 2009;4:12.

[18] Lara B, Martínez-Delgado B, Torres ML, Marín-Arguedas S, Bustamante A, Miravitlles M. Alpha-1-antitrypsin deficiency associated with the Mattawa variant. Arch Bronconeumol 2013;49:548–50.

[19] Joly P, Guillaud O, Hervieu V, Francina A, Mornex JF, Chapuis-Cellier C, et al. Clinical heterogeneity and potential high pathogenicity of the Mmalton alpha 1 antitrypsin allele at the homozygous, compound heterozygous and heterozygous states. Orphanet J Rare Dis 2015;10:130.

[20] Curiel DT, Vogelmeier C, Hubbard RC, Stier LE, Crystal RG. Molecular basis of alpha 1-antitrypsin deficiency and emphysema associated with the alpha 1-antitrypsin Mmineral springs allele. Mol Cell Biol 1990;10:47–56.

[21] Matsunaga E, Shiokawa S, Nakamura H, Maruyama T, Tsuda K, Fukumaki Y. Molecular analysis of the gene of the alpha 1-antitrypsin deficiency variant, Mnichinan. Am J Hum Genet 1990;46:602–12.

[22] Faber J-P, Poller W, Weidinger S, Kirchgesser M, Schwaab R, Bidlingmaier F, et al. Identification and DNA sequence analysis of 15 new alpha-1 antitrypsin variants, including 2 PiQO alleles and one deficient PiM allele. Am J Hum Genet 1994;55:1113–21.

[23] Takahashi H, Nukiwa T, Satoh K, Ogushi F, Brantly M, Fells G, et al. Characterization of the gene and protein of the alpha 1-antitrypsin "deficiency" allele Mprocida. J Biol Chem 1988;263:15528–34.

[24] Jardi R, Rodriguez-Frias F, Lopez-Talavera JC, Miravitlles M, Cotrina M, Costa X, et al. Characterization of the new alpha-1-antitrypsin-deficient PI M-type allele, PI M(vall d'hebron) (Pro(369)--. Ser). Hum Hered 2000;50:320–1.

[25] Zorzetto M, Russi E, Senn O, Imboden M, Ferrarotti I, Tinelli C, et al. SERPINA1 gene variants in individuals from the general population with reduced alpha1-antitrypsin concentrations. Clin Chem 2008;54:1331–8.

[26] Coni P, Pili E, Convertino G, Pichiri G, Balestrino A, Delmastro M, et al. MVarallo: a new M(Like) alpha 1-antitrypsin-deficient allele. Diagn Mol Pathol 2003;12:237–9.

[27] Owen MC, Brennan SO, Lewis JH, Carrell RW. Mutation of antitrypsin to antithrombin. Alpha 1-antitrypsin Pittsburgh (358 Met leads to Arg), a fatal bleeding disorder. N Engl J Med 1983;309:694–8.

[28] Faber JP, Weidinger S, Goedde HW, Ole K. The deficient alpha-I-antitrypsin phenotype PI P is associated with an A-to-T transversion in exon III of the gene. Am J Hum Genet 1989;45: 161–3.

[29] Hildesheim J, Kinsley G, Bissell M, Pierce J, Brantly M. Genetic diversity from a limited repertoire of mutations on different common allelic backgrounds: alpha 1-antitrypsin deficiency variant Pduarte. Hum Mutat 1993;2:221–8.

[30] Holmes MD, Brantly ML, Crystal RG. Molecular analysis of the heterogeneity among the P-family of alpha-1-antitrypsin alleles. Am Rev Respir Dis 1990;142:1185–92.

[31] Yuasa I, Sugimoto Y, Ichinose M, Matsumoto Y, Fukumaki Y, Sasaki T, et al. PI*S(iiyama), a deficiency gene of alpha 1-antitrypsin: evidence for the occurrence in western Japan. Jpn J Hum Genet 1993;38:185–91.

[32] Seyama K, Nukiwa T, Souma S, Shimizu K, Kira S. Alpha 1-antitrypsin-deficient variant Siiyama (Ser53[TCC] to Phe53[TTC]) is prevalent in Japan. Status of alpha 1-antitrypsin deficiency in Japan. Am J Respir Crit Care Med 1995;152:2119–26.

[33] Holmes MD, Brantly ML, Fells GA, Crystal RG. Alpha 1-antitrypsin Wbethesda: molecular basis of an unusual alpha 1-antitrypsin deficiency variant. Biochem Biophys Res Commun 1990;170:1013–20.

[34] Jardi R, Rodriguez F, Miravitlles M, et al. Identification and molecular characterization of the new alpha-1-antitrypsin deficient allele PI Y barcelona (Asp256--. Val and Pro391--. His). Mutations in brief no. 174. Online. Hum Mutat 1998;12:213.

[35] Weidinger S, Jahn W, Cujnik F, Schwarzfischer F. Alpha-1-antitrypsin: evidence for a fifth PI M subtype and a new deficiency allele PI*Z augsburg. Hum Genet 1985;71:27–9.

[36] Whitehouse DB, Abbott CM, Lovegrove JU, McIntosh I, McMahon CJ, Mieli-Vergani G, et al. Genetic studies on a new deficiency gene (PI*Ztun) at the PI locus. J Med Genet 1989;26:744–9.

[37] American Thoracic Society/European Respiratory Society. American Thoracic Society/European Respiratory Society Statement: standards for the diagnosis and management of individuals with alpha1-antitrypsin deficiency. Am J Respir Crit Care Med 2003;168:818–900.

[38] Lomas DA, Evans DL, Finch JT, Carrell RW. The mechanism of Z alpha 1-antitrypsin accumulation in the liver. Nature 1992;357:605–7.

[39] Teckman JH, An JK, Blomenkamp K, Schmidt B, Perlmutter D. Mitochondrial autophagy and injury in the liver in alpha 1-antitrypsin deficiency. Am J Physiol Gastrointest Liver Physiol 2004;286:851–62.

[40] Ogushi F, Fells GA, Hubbard RC, Straus SD, Crystal RG. Z-type alpha 1-antitrypsin is less competent than M1-type alpha 1-antitrypsin as an inhibitor of neutrophil elastase. J Clin Invest 1987;80:1366–74.

[41] Engh R, Löbermann H, Schneider M, Wiegand G, Huber R, Laurell CB. The S variant of human alpha 1-antitrypsin, structure and implications for function and metabolism. Protein Eng 1989;2:407–15.

[42] Tan L, Dickens JA, Demeo DL, Miranda E, Perez J, Rashid ST, et al. Circulating polymers in α1-antitrypsin deficiency. Eur Respir J 2014;43:1501–4.

[43] Teckman JH. Liver disease in alpha-1 antitrypsin deficiency: current understanding and future therapy. COPD 2013;10(Suppl. 1):35–43.

[44] Pan S, Huang L, McPherson J, Muzny D, Rouhani F, Brantly M, et al. Single nucleotide polymorphism-mediated translational suppression of endoplasmic reticulum mannosidase I modifies the onset of end-stage liver disease in alpha1-antitrypsin deficiency. Hepatology 2009;50:275–81.

[45] Lam M, Torbenson M, Yeh MM, Vivekanandan P, Ferrell L. HFE mutations in alpha-1-antitrypsin deficiency: an examination of cirrhotic explants. Mod Pathol 2010;23:637–43.

Chapter 6

Alpha-1 Antitrypsin Deficiency: Lung Pathophysiology

6.1 INTRODUCTION

Chronic obstructive pulmonary disease (COPD) is characterized by a generalized airways inflammation with progressive airflow obstruction only partly reversible and systemic comorbidities [1]. Although smoking is the main factor of risk for COPD, only a minority of smokers (approximately 15%–20%) develops this disease. COPD is associated with an enhanced chronic inflammatory response in the airways and the lungs to inhaled cigarette smoke and other noxious particles, such as occupational exposures to dusts, chemical agents, fumes, and smoke from biomass fuels. This chronic inflammatory response may induce parenchymal tissue destruction resulting in emphysema, and small airway stenosis/occlusion due to fibrosis [2]. The pathophysiological mechanisms for this amplified inflammation are not totally understood, but several abnormal characteristics that may have an important role on its development and maintenance have been identified, including: (1) a specific inflammatory cell pattern in the airways, lung parenchyma, and pulmonary vasculature, involving increased numbers of CD8+ Tc1 lymphocytes, neutrophils, and macrophages; (2) a protease–antiprotease imbalance caused by an excess of proteases released by inflammatory cells and lung epithelial cells; (3) a disturbance in the balance between the production of antioxidant defenses and a excessive production of reactive oxygen species (ROS) (generated by cigarette smoke and other inhaled particulates, and oxidants released from neutrophils and macrophages); (4) a variety of chemotactic factors attracting inflammatory cells from the circulation, as well as proinflammatory cytokines and growth factors inducing structural changes; (5) a presence of autoantigens inducing autoimmunity [3]; and (6) a chronic colonization of the lower respiratory tract by bacterial pathogens inducing a chronic inflammatory response with lung damage [4].

6.2 PHYSIOPATHOLOGICAL MECHANISMS OF ALPHA-1 ANTITRYPSIN DEFICIENCY-RELATED COPD

A significant familial risk of airflow limitation has been observed in smoking siblings of patients with severe COPD [5], this fact strongly suggesting that genetic together with environmental factors may influence this susceptibility. Although several genome-wide association studies indicate a possible role of other genes, there remains a discrepancy between genome-wide association study analyses and candidate gene analyses [5], and really the only well documented genetic factor of risk for COPD so far is severe hereditary alpha-1 antitrypsin (AAT) deficiency (AATD), which is behind of at least 1%–3% of all cases of COPD [6,7].

6.2.1 Alpha-1 Antitrypsin Genotypes Related to COPD Development

Emphysema associated with the inherited serum AATD appears to result from an imbalance between neutrophil elastase and its major inhibitor (i.e., human AAT) within the alveolar structures (Fig. 6.1). In fact, neutrophil elastase is the most recognized etiologic factor in COPD associated with AATD. The serum level of AAT to protect the lungs against free neutrophil elastase is based on an observational study conducted 20 years ago by Turino et al., who established the cutoff in 11 micromole per liter (μM/L), measured by the NHLBI Registry USA standard [8]. This value is equivalent to approximately 35% of the mean plasma

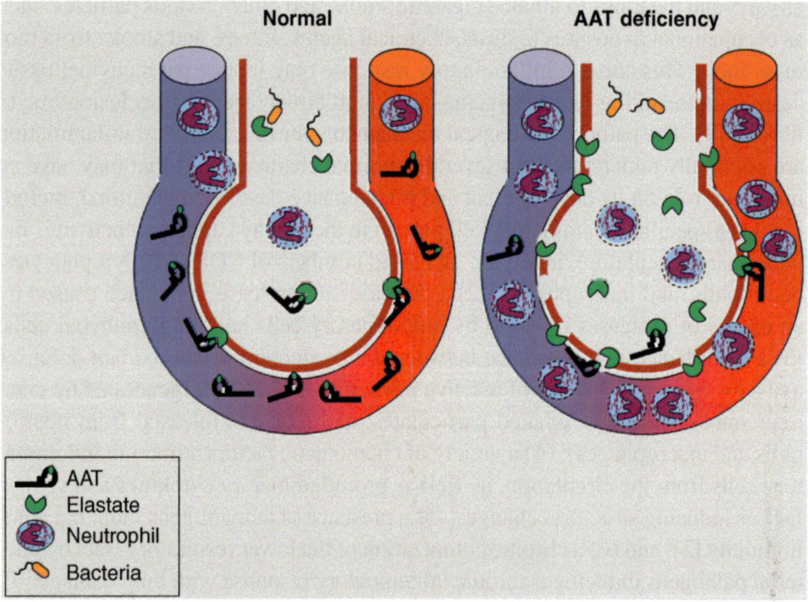

FIGURE 6.1 Classical hypothesis of the alpha-1 antitrypsin *(AAT)* deficiency mechanism of COPD development. Under normal conditions there is a balance between AAT and neutrophil elastase liberated to defend the human body from invading bacteria. The AAT deficiency (AATD)-related emphysema is characterized by increased levels of free elastase and a high neutrophil burden into the airways of AATD individuals, resulting in a major imbalance in favor of elastase regarding its inhibitor (AAT) within the bronchioles and alveolar structures. In these circumstances, elastase can degrade the walls of the alveoli and bronchi, causing lung emphysema, chronic bronchitis, and bronchiectasis in the long term.

AAT predicted value, or 57–60 mg/dL (if it is determined by nephelometry), or 80 mg/dL (if the measurement is made by radial immunodiffusion, a technique in disuse). In clinical practice, the vast majority of cases with AATD-related COPD are homozygous ZZ, only a small number are heterozygous SZ, and very infrequently homozygous or compound heterozygous of Z, rare or null alleles [7]. Rare deficiency genotypes have already been mentioned in Chapter 5 and are listed in Table 5.1.

The introduction of genotyping with PCR in recent decades has led to the identification of more than 30 new AATD null (also called QO) variants, which have been characterized in individuals of all races, but predominantly in Caucasians [9–34]. The designation of these variants is justified by the fact that they virtually express undetectable amounts (<1%) of plasma AAT (Table 6.1). Null variants are caused by nucleotide changes in the AAT molecule, deletions, or additional insertions of nitrogenous bases in both the coding (exons II–V) or the promoter (IA–IC) regions of the gene, generating stop codons, affecting the splicing of mRNA, or translating nonfunctional, truncated or unstable proteins that are degraded within the synthesizing cells without being secreted into the blood. There is not enough information on its prevalence, but they are very rare. However, despite its rarity they should be taken into consideration because of the high risk involved for emphysema development. QO genotypes do not cause liver disease because by not producing proteins they do not generate cell stress in hepatocytes nor liver damage [7].

6.2.2 Mechanisms

It is widely accepted that chronic inflammation plays a key role in the patho physiology of COPD associated with hereditary AATD, and it is assumed that the basic pathophysiological mechanisms should be the same (though amplified) that lead to common COPD. However, the profound mechanisms of lung injury in AATD are complex and not completely known [7]. Although once started the inflammatory process several proinflammatory factors can act synergistically, for didactic reasons they are described separately in the subsequent sections.

6.2.2.1 Protease-Mediated Lung Disease

The first pathophysiological event which leads to pulmonary emphysema in Pi*ZZ genotype-related AATD is the retention of Z AAT in the liver, secondarily accompanied by a directly proportional decrease of AAT concentration in serum and tissues. This decrease in blood concentration together with the significantly reduced inhibitory capacity of the Z-AAT protein generates a protease–antiprotease imbalance in the lungs, with overexpression of free neutrophil proteases and activated neutrophils (more pronounced in respiratory exacerbations), playing a key pathophysiological role in the initiation, development, and

TABLE 6.1 Alpha-1 Antitrypsin Deficiency Null (QO) variants (Sorted by Alphabetical Order)

Allele	Mutation	Clinical details	Author, country, and year [references]
QOamersfoort	Exon II. Tyr160-stop. Nonsense mutation generating a stop codon at 160, not detectable mRNA	A Caucasian woman with COPD	Prins et al., The Netherlands 2008 [9]
QObellingham	Exon III. Stop codon Lys 217 which prevents mRNA translation	An index case with emphysema	Satoh et al., USA 1988 [10]
QObolton	Exon V. Deletion Pro362, generating a stop codon at 373. Translation of a truncated protein which is degraded intracellularly	An isolated case. High risk of emphysema in homozygotes and compound heterozygotes	Fraizer et al., Canada 1989 [11]
QObonny blue	Intron II. Deletion of G at position 1, affecting mRNA splicing	An isolated case. High risk in homozygotes and compound heterozygotes	Lee and Brantly, USA 2000 [12]
QObrescia	T for G transversion (GAG→TAG) at codon 257, which generates a stop codon	Two homozygotes and one heterozygous Z/QObrescia (all smokers) with emphysema, and three healthy M/QObrescia	Ferrarotti et al., Italy 2014 [13]
QOcairo	Exon III. Lys 259 stop codon. Translation of a truncated protein which is degraded intracellularly	Detected in three subjects: an Egyptian/Italian descent and two Italians from southern Italy	Zorzetto et al., Italy 2005 [14]
QOcork	Exon II. Deletion in codon 189 CAG, which generates a premature stop codon	An Irish woman M1/QOcork with asthma	Ferrarotti et al., Italy 2014 [13]
QOcosenza	Exon IV. C by T transversion in codon 305, which generates a premature stop codon	A family with two M/QOcosenza and two S/QOcosenza subjects	Ferrarotti et al., Italy 2014 [13]

TABLE 6.1 **Alpha-1 Antitrypsin Deficiency Null (QO) variants (Sorted by Alphabetical Order)** (*cont.*)

Allele	Mutation	Clinical details	Author, country, and year [references]
QOclayton	Exon V. Insertion of a C next to Pro362, generating a stop codon at 373	An isolated case. High risk in homozygotes and compound heterozygotes	Brantly et al., USA 1997 [15]
QOdevon (=QOnewport)	Exons II and V. Gly115Ser and Glu342Lys (Z mutation). The mutant protein is degraded into hepatocyte	Isolated cases. It is unclear whether the Gly115Ser mutation may favor the development of disease in absence of Z mutation	Graham et al., UK 1990 [16]
QOdublin	T deletion and shift of the reading frame generating a stop codon at TAA 373. No detectable AAT	Reported four Irish M1/QOdublin: a woman with bronchiectasis (index case), and three healthy children	Ferrarotti et al., Ireland 2014 [13]
QOgranite falls	Exon II. Tyr160 1 bp deletion that generates a stop codon. No detectable mRNA	Severe emphysema in an African-American compound heterozygote (Z/QO)	Nukiwa et al., USA 1987 [17]
QOhong kong	Exon IV. Deletion Leu318 and stop codon at 334. Truncated protein which is degraded intracellularly	Isolated cases in subjects with Chinese ancestry. High risk in homozygotes and compound heterozygotes	Sifers et al., EEUU 1988 [18]
QOisola di procida	Extensive 17 KB deletion (exons II–V). Undetectable mRNA and absence of protein in blood	An heterozygote QOisola procida/ Mprocida	Takahashi and Crystal, USA 1990 [19]
QOkayseri	Exon II. Insertion of C and A with reading frame shift at codon 73, generating a truncated protein lacking of active site	A QOkayseri homozygote Turkish woman from Kayseri, resident in Munich, with diffuse cystic bronquiectasias (without emphysema) and pulmonary embolism	Milger, Germany 2015 [20]

(*Continued*)

TABLE 6.1 **Alpha-1 Antitrypsin Deficiency Null (QO) variants (Sorted by Alphabetical Order)** *(cont.)*

Allele	Mutation	Clinical details	Author, country, and year [references]
QOlampedusa	Exon V. Deletion of G at 337 GTG codon, generating an adjacent stop codon (TGA 338)	Six heterozygotes M/QOlampedusa: two with emphysema (no smokers), two chronic bronchitis (one smoker) and two healthy (one smoker)	Ferrarotti et al., Italy 2014 [13]
QOlisbon	Exon II. Thr68Ile. Translation of a truncated protein, which is degraded into the hepatocyte	A heterozygote M/QO Lisbon, expressing 50% of serum AAT	Faber et al., Germany 1994 [21]
QOludwigshafen	Exon II. Ile92Asn. Great alteration of the protein structure, which is degraded without being secreted	Isolated cases. High risk in homozygotes and compound heterozygotes	Faber et al., Germanya 1994 [21]; Frazier et al., Canada 1990 [22]
QOmadrid	Duplication of T in exon 1C, that interrupts the splicing of the mRNA in hepatocytes	Two brothers QOporto/QOmadrid with severe AAT deficiency (the index case with emphysema)	Lara et al., Spain 2014 [23]
QOmattawa	Exon V. Leu353Phe (M1), generating a stop codon at 376	Isolated cases. High risk in homozygotes and compound heterozygotes	Curiel et al., USA 1989 [24]; Lara et al., Spain 2013 [25]
QOourém	Exon V. Identical to QOmattawa but in M3 (Leu353Phe, stop codon 376). Intracell degraded protein	Isolated cases (41 from Portugal). High risk in homozygotes and compound heterozygotes	Seixas et al., Portugal 2002 [26]; Vaz et al., Portugal 2012 [27] Hernández, Spain 2014 [28]
QOmilano	Exon III. Deletion of 17 bp, generating a stop codon downstream. Truncated protein nonfunctional due to the lack of active site	Described in an Italian boy M/QOmilano with moderately low serum AAT levels	Rametta et al., Italy 2013 [29]

TABLE 6.1 Alpha-1 Antitrypsin Deficiency Null (QO) variants (Sorted by Alphabetical Order) ***(cont.)***

Allele	Mutation	Clinical details	Author, country, and year [references]
QOnew hope	Exons IV and V. Double mutation: Gly320Glu and Glu342Lys (=Z mutation)	Isolated case. High risk in homozygotes and compound heterozygotes	Brantly et al., USA 1992 [30]; Brantly et al., USA 1996 [31]
QOperugia	Exon III. Deletion of G at GTG 239 codon, generating a stop codon at TGA 241	A 50-year-old smoker with emphysema. A healthy sibling with the same mutation	Ferrarotti et al., Italy 2014 [13]
QOpordenone	Exon IV. Deletion of a T at TGA 338 generating a distal stop codon	Ten cases M/ QOpordenone, two with emphysema; but the remaining eight healthy	Ferrarotti et al., Italy 2014 [13]
QOporto	Exon IA. Absence of normal splicing between exon IA and II	A heterozygote M/ QOporto, expressing 50% of serum AAT	Seixas et al., Portugal 2002 [26]
QOtorino	Exon IV. T by A transversion in codon 297 (TyrTAT > TerTAA)	Reported a case S/QOtorino in a smoker with emphysema	Ferrarotti et al., Italy 2014 [13]
QOriedenburg	Exon IC. Extensive deletion above the promoter exon IC. No protein expression	Isolated cases. High risk in homozygotes and compound heterozygotes	Poller et al., Germany 1991 [32]; Faber et al., Germany 1994 [21]
QOsaarbrucken	Exon V. Insertion of cytosine (C) in Pro362, generating a stop codon at 376	Reported a heterozygote M/ QOsaarbueken expressing 50% of serum AAT	Faber et al., Germany 1994 [21]
QOsoest	Exon II. Deletion of A at 102 codon (ACC→ −CC) with stop codon at 112. No protein secretion	An index case QOsoest/Pi*Z with emphysema	Prins et al., The Netherlands 2008 [9]

(Continued)

TABLE 6.1 Alpha-1 Antitrypsin Deficiency Null (QO) variants (Sorted by Alphabetical Order) (*cont.*)

Allele	Mutation	Clinical details	Author, country, and year [references]
QOtrastevere	Exon III. Tyr194 stop codon. Truncated protein scantily secreted	A compound heterozygote Mprocida/ QOtrastevere with emphysema	Lee et al., USA 1998 [33]
QOwest	Intron II. Change of G by T with abnormal splicing of premessenger RNA	A compound heterozygote with emphysema	Laubach et al., USA 1993 [34]

A, Adenine; Ala, alanine; Arg, arginine; Asn, asparagine; Asp, aspartic acid; bp, base pairs; COPD, chronic obstructive pulmonary disease; Cys, cysteine; C, cytosine; G, guanine; Glu, glutamic acid; Gly, glycine; His, histidine; Ile, isoleucine; KB, unit for size of nucleic acids corresponding to a length of 1000 nucleotides (=thousand bp); Leu, leucine; Lys, lysine; Met, methionine; mRNA, messenger RNA; T, thymine; Phe, phenylalanine; Pro, proline; Ser, serine; Thr, threonine; UK, United Kingdom; USA, United States of America; Val, valine.

maintenance of a chronic inflammatory destruction–repair process that eventually lead to a irreversible destruction of alveoli (Fig. 6.1).

The hypothesis of the imbalance between AAT and neutrophil elastase began to develop more than 50 years ago. In fact, approximately coinciding with the discovery by Laurell and Eriksson of AATD in Malmoe in the early 1960s of last century, Gross et al. in Pittsburg discovered that intratracheal instillation of papain caused lung emphysema in both normal and silicotic rats with silica-induced pulmonary fibrosis [35].

Three years after (1967) Janoff and Scherer discovered a powerful elastolytic protease in polymorphonuclear leukocytes, which they called human neutrophil elastase [36].

In 1969, Turino et al. reported for the first time that the plasma of patients with AATD-related COPD was not able to properly neutralize pancreatic elastase, and advanced the likely role of elastase in the pathogenesis of COPD [37].

In 1972, Janoff discovered that circulating AAT effectively inhibited human neutrophil elastase [38], and in subsequent years other authors showed the capacity of several kinds of proteases to degrade elastin, causing emphysema when instilled into the lungs of healthy rats, mice, hamsters, guinea pigs, and dogs [39–43].

In 1977, Janoff et al. reported that human leukocyte elastase penetrated the dog alveolar septal connective tissue after its airway instillation, and that the enzyme attached to elastic fibers, inducing histologic changes comparable to those seen in human emphysema [44].

In addition, between 1977 and 1980, Janoff and other researchers showed that the oxidation of AAT by cigarette smoke inactivated the AAT inhibitory capacity of elastase [45–48].

All these findings together lead to the hypothesis that pulmonary emphysema was due to the proteolytic action of neutrophil proteases on lung elastic tissue by a congenital lack of AAT (hereditary deficiency) or due to its inactivation by cigarette smoke ("acquired" deficiency) [49,50]. Currently, this hypothesis still remains valid, but it has been enriched with new scientific contributions, that have shown that the interaction between different types of proteases is essential for activating a complex proteolytic cascade, which plays an essential role in the emphysema pathophysiology.

In the late 1980s, it was demonstrated that cathepsin G and proteinase-3 (two serine proteases stored with elastase in the azurophilic granules of neutrophil) when instilled into the trachea of hamsters produced bronchial metaplasia and emphysema in these animals [51,52]. Likewise, intratracheal instillation of cathepsin B (a lysosomal cysteine protease of the papain family) inducted lung emphysema in hamsters [53].

In this context, recently it has been found that the proteinase-3, accumulated in the azurophilic granules of neutrophil, upon its release promotes itself apoptosis of endothelial cells, and may play an important pathogenic role in lung injury for its high concentration in neutrophils, its lowest affinity—compared with elastase—by AAT, its greater tendency to join the alpha-2-macroglobulin (for which it is retained without being neutralized), and its greater penetration into the interstitium after neutrophil degranulation [54].

Neutrophil elastase has been situated diagrammatically at the apex of a hierarchical tree of proteases, acting as the principal regulator of several classes of tissue-degrading proteases. For example, cleavage of the zymogen procathepsin B from bronchial epithelial cells by neutrophil elastase generates cathepsin B activity [55]. Besides, free neutrophil elastase can activate several tissue metalloproteinases (MMPs), such as MMP-2 (gelatinase A), MMP-3 (matrix metallopeptidase 3) [56], or MMP-9 (gelatinase B) [57]. Furthermore, in cell cultures and in a neutrophil elastase knockout mice model, neutrophil elastase upregulated both cathepsin B and matrix metalloprotease-2 expression [58].

In humans, bronchoalveolar lavage (BAL) fluid from patients with pneumonia and those with AATD compared with healthy volunteers, free neutrophil elastase increased cathepsin B and MMP-2 activities. The addition of AAT to the patients BAL fluid greatly reduced elastase-induced cathepsin B and MMP-2 expression in vitro, and AAT augmentation therapy to AATD individuals also reduced cathepsin B and MMP-2 activity in BAL fluid in vivo [59].

Some studies have revealed that human neutrophil elastase, via activation of the Hageman factor-prekallikrein system, or directly, releases kinin from kininogens, and these peptide exerts various biological activities, such as vascular permeability increase and smooth muscle contraction, although the practical impact of this activity is unknown [60].

Neutrophil elastase can degrade most extracellular matrix and plasma proteins, such as serpins, polypeptides, the surfactant proteins SP-A and D, lactoferrin, immunoglobulins, and the C3bi complement fraction [61] (Fig. 6.2).

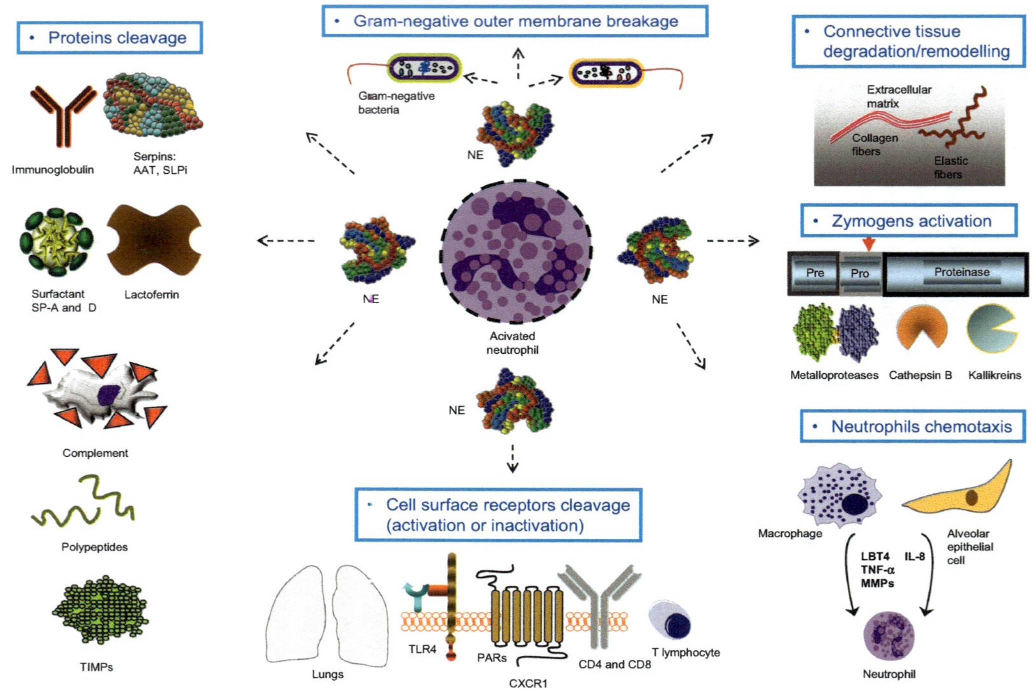

FIGURE 6.2 **Physiological functions and destructive effects of an excess of free neutrophil elastase.** The main physiological functions of neutrophil elastase are the degradation in lysosomes of foreign organic molecules, the lysis of the cell wall of invading bacteria, and the proteolysis of elastin and collagen-IV for extracellular matrix remodeling. An excess of unneutralized free elastase can degrade almost all extracellular matrix proteins, and several other proteins, such as serpins, TIMPs, surfactant proteins, lactoferrin, immunoglobulins, and complement. Elastase can also cleave and activate metalloproteases, cathepsin B, and kallikrein proenzymes. Acting on several membrane receptors, such as CD4, CD8, toll-like, PARs, or CXCR1, significantly alters the inflammatory response and activity of lymphocytes, neutrophils, and macrophages. *AAT*, Alpha-1 antitrypsin; *CD4 and CD8*, CD4 lymphocytes (T helper) and CD8 (T cytotoxic); *CXCR1*, chemokine (C-X-C motif) receptor 1 (interleukin 8 receptor); *IL-8*, interleukin 8; *LTB-4*, leukotriene B-4; *MMPs*, metalloproteinases; *NE*, neutrophil elastase; *PARs*, proteinase-activated receptors; *SLPi*, secretory leukocyte protease inhibitor; *TIMPs*, tissue inhibitors of metalloproteinases; *TLR*, toll-like receptor; *TNF-α*, tumor necrosis alpha.

Remarkably, neutrophil elastase inactivates the TIMPs, as it has been demonstrated in a study founding that it participated in the connective tissue destruction not only by its direct action on matrix macromolecules but also by a specific inactivation of TIMP-1 while activating the conversion of the zymogen pro-MMP-3 in active MMP-3 [62].

On the other hand, neutrophil proteinases released at sites of inflammation can affect tissue function by either activating or disarming the signal transduction mediated by PARs, a subfamily of G protein-coupled receptors, numbered from one to four, which are expressed by variety of cells, including leukocytes, epithelial cells, endothelial cells, or fibroblast [63,64]. In addition, neutrophil elastase cleaves the lymphocyte surface antigens CD2, CD4, and CD8 on T lymphocytes, leading to a significant reduction of their cytotoxicity toward target cells, and IL-2 and IL-4 production [65].

In BAL from AATD patients, it has been found threefold increase in neutrophil chemotactic activity, which is normalized with AAT therapy [66].

Neutrophil elastase contributes to the chemotaxis of neutrophils by stimulating leukotriene B4 (LTB4) production by macrophages, a potent chemoattractant of neutrophils able to induce adhesion and activation of leukocytes on endothelium, allowing them to bind to and cross it into the inflamed tissues. Neutrophil recruitment to the airways is an important component of continuing inflammation and progression of COPD, particularly in presence of AATD. Nearly 20 years ago, a pioneering study showed higher levels of LTB4 and chemotactic activity in alveolar macrophage supernatant obtained from BAL fluid of AATD patients compared with nondeficient subjects. In this study, purified human AAT was able to prevent neutrophil elastase to release LTB4 from stimulating the macrophages to release the chemotactic factor [67].

Later it was found that bronchial secretions of COPD patients with AATD had increased neutrophil chemotactic activity, attributable to increased levels of interleukin 8 (IL-8) and LTB4. These data led to the conclusion that the absence of a normal antineutrophil elastase screen in the lower respiratory tract permits free neutrophil elastase to bind to alveolar macrophages, resulting in the release of LTB4, a process which attracts neutrophils to the alveoli accelerating the lung destruction that characterizes this disorder [68].

Recently, in vitro studies demonstrated that neutrophil elastase stimulates both LTB4 production and the circulating leukocytes membrane receptor LTB4 1 (BLTR1) expression. Interestingly, AAT can bind LTB4 decreasing plasma LTB4 concentrations and membrane bound of elastase [69].

In addition, neutrophil elastase upregulates IL-8 gene expression in human bronchial epithelial cells, via the NF-κB pathway, through the cell surface membrane-bound toll-like receptor 4 (TLR-4) [70,71]. IL-8 is a chemokine produced by macrophages, and other cell types, such as epithelial cells, airway smooth muscle cells, and endothelial cells. IL-8 induces neutrophil chemotaxis and phagocytosis, via the neutrophils membrane G protein-coupled receptors CXCR1 and CXCR2 (Fig. 6.2). In vitro, AAT can bind to IL-8 (a ligand for CXCR1) and the AAT–IL-8 complex formation prevents IL-8 interaction with CXCR1 [66].

6.2.2.2 Oxidant-to-Antioxidant Imbalance

Oxidative stress is involved in the pathogenesis of airway obstruction in COPD, and also plays an important role in the airway obstruction of AATD patients [72]. Excess oxidants are generated either exogenously from cigarette smoke or air pollutants, or endogenously from activated neutrophils and macrophages releasing ROS, leading to an oxidant-to-antioxidant imbalance in the lungs which causes oxidative inactivation of antiproteases, surfactants, mucus hypersecretion, membrane lipid peroxidation, alveolar epithelial injury, apoptosis, remodeling of extracellular matrix, nucleic acids (especially RNA) damage, and increased gene expression of proinflammatory mediators [73].

6.2.2.3 Tumor Necrosis Factor-Related Impaired Neutrophil Activity

Tumor necrosis factor alpha (TNF-α) signaling is a major pathogenic player in the pathogenesis of both genetic AATD and nongenetic COPD [74]. Ex vivo studies have shown elevated levels of TNF-α in peripheral blood, sputum, bronchial biopsies, and BAL fluid from patients with stable COPD and during acute exacerbations [75], and cell culture experiments have demonstrated that TNF-α can enhance chemotaxis and cell migration by inducing the expression of IL-8 in alveolar epithelial cells while upregulating ICAM-1, E-selectin, and VCAM-1 endothelial adhesion molecules in sites of inflammation and tissue injury [76,77].

Neutrophil degranulation and apoptosis is increased nearly twofold in Pi*ZZ individuals compared with healthy controls, due to an abnormal activation of the TNF-α system [78]. Contact with pollutants including cigarette smoke promotes TNF-α production by alveolar macrophages, neutrophils, T cells, mast cells, and epithelial cells. Moreover, in AATD, misfolded AAT proteins accumulate in the endoplasmic reticulum (ER) of neutrophils (similarly to as it occurs in hepatocytes), leading to ER stress with overexpression of TNF-α proapoptotic signals and activation of the membrane-anchored protein ADAM-17 (or TACE, *tumor necrosis factor-α-converting enzyme*), normally inhibited by AAT, and thereby increasing the shedding of soluble TNF-α. The end result of these pathological processes is an increased neutrophil apoptosis and defective bacterial killing. Interestingly, treatment with AAT augmentation therapy decreases neutrophil ADAM-17 activity and apoptosis in vivo and increases bacterial killing, breaking the TNF-α-induced inflammation cycle by binding TNF1 and TNF2 receptors (TNFR1/2) avoiding the activation of the transcription factor NF-κB which mediates apoptosis, and modulating TNF-α receptor engagement [75,78].

Further evidence demonstrating that AATD patients are characterized by increased activation of the TNF-α system has been the finding of an increased membrane TNF-α levels, and plasma concentrations of TNFR1 and neutrophil-released secondary and tertiary granule antigenic proteins in AATD patients.

In addition, the incidence of autoantibodies directed against degranulated lactoferrin and surface protein accessible to these antibodies was increased in ZZ-AATD subjects, leading to an enhanced rate of neutrophil ROS production. Interestingly, in these experiments AAT augmentation therapy resulted in a decreased membrane TNF-α expression and plasma levels of granule antigenic proteins and immunoglobulin G class antilactoferrin autoantibodies [75].

6.2.2.4 Autoimmunity-Mediated COPD

Within the lungs of patients with severe emphysema, inflammation continues despite smoking cessation; and breakdown of the extracellular matrix, particularly elastin, with antielastin antibody and T-helper type 1 responses, which can perpetuate and/or aggravate lung destruction, are a key feature of both "common" COPD and AATD-related COPD [79,80]. Besides, it has been demonstrated that IgG autoantibodies with avidity for pulmonary epithelium are prevalent in patients with COPD [81], and evidences from animal models also supports a role for autoimmunity in COPD [82,83], even though other researchers have reported that neither antielastin nor anti-*N*-acetylated proline-glycine-proline autoantibodies are objectively evident in some COPD subjects [84,85].

In lungs from COPD and emphysema patients removed surgically a lymphocytic infiltrate was found with increasing pulmonary CD4+ and CD8+ T-cell numbers relating to worsening disease [86,87], and more recently, in explanted lungs from patients with usual COPD and AATD-related COPD an important adaptive immune inflammation comprising B, CD4+, and CD8+ lymphocytes, and lymphoid follicles have been found, indicating that a complex mechanism involving the adaptive immune system occurs in AATD similarly to what occurs in usual COPD [88].

Therefore the participation of immune mechanisms in AATD-related COPD is still under discussion, and although it is generally recognized that there may be an aberrant immune response in AATD-related COPD patients, its practical impact is doubted by some experts, being precise more well-designed and powerful studies to clarify this question [89].

6.2.2.5 Accumulation of Z Protein Polymers in the Airways

In the last decade it has been demonstrated the formation of Z-AAT polymers into alveolar epithelial lung cells and the existence of polymers colocalized with neutrophils in the lung interstitium. Intracellular polymers, just like in hepatocytes, cause ER stress and proapoptotic epithelial lung cellular dysfunction, while interstitial polymers have a neutrophil chemoattractant activity, which increases neutrophilic inflammation and amplifies inflammation [90–95].

More recently, it has been shown that intracellular Z-AAT polymers are also formed into the bronchial epithelial cells of Pi*ZZ subjects and this fact can contribute to increasing inflammation of the airways [96].

6.3 CLINICAL EXPRESSION

The development of COPD in the setting of severe AATD is highly variable, since AATD is a complex monogenic disorder with a great variability in the disease presentation, and it is not known what percentage of AATD subjects develops COPD and what does not. Although the reason for this variability is unknown, it is recognized that AATD is not a disease but a condition that predisposes to the development of diseases, being generally necessary that concur in a same person, in addition to the congenital deficiency of AAT, other aggravating environmental factors, and/or other modifier genes not well known (Fig. 6.3).

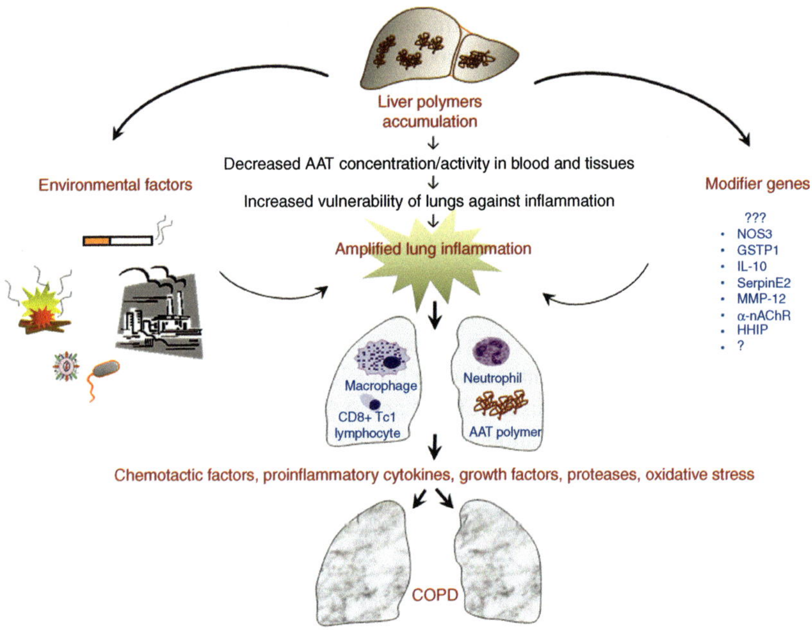

FIGURE 6.3 Mechanisms of COPD associated with alpha-1 antitrypsin deficiency. The typical pathophysiological event in Pi*Z homozygotes is the retention of Z AAT in the liver, accompanied by a proportional decrease of the AAT concentration in serum and tissues, which together with its significantly reduced inhibitory capacity, generates a protease–antiprotease imbalance in the lungs. COPD is characterized by a specific pattern of inflammation involving increased numbers of CD8+ Tc1 lymphocytes, neutrophils, and macrophages, which release inflammatory mediators and enzymes, and interact with structural cells in the airways, lung parenchyma, and pulmonary vasculature. The tissue damage caused by these agents is further potentiated by the mutual interactions between the protease/antiprotease imbalance, autoimmunity, apoptosis, oxidative stress, lung polymers, proinflammatory cytokines, and growth factors, leading to the development and maintenance of a chronic inflammatory destruction–repair process that would eventually lead to the irreversible destruction of bronchial walls and alveoli. *AAT*, Alpha-1 antitrypsin; *α-nAChR*, alpha-nicotinic acetylcholine receptor; *CD8*, cluster of differentiation 8; *CD8+ Tc1*, cluster of differentiation 8 positive T-cell cytotoxic type 1; *GSTP1*, glutathione S-transferase P; *HHIP*, hedge-hog interacting protein; IL-10, interleukin-10; *MMP-12*, matrix metalloproteinase-12, also known as macrophage metalloelastase (MME) or macrophage elastase (ME); *NOS3*, nitric oxide synthase 3; *SerpinE2*, serpin peptidase inhibitor, clade E (nexin, plasminogen activator inhibitor type 1), member 2.

It is beyond question that smoking is a triggering factor highly relevant in the development and maintenance of COPD [7,97]. The importance of smoking was reflected by the fact that 90% of Pi*ZZ smokers developed emphysema, while between 30% and 40% of nonsmokers did not [98]. In addition, industrial and agricultural pollution, and indoor smoke from burning biomass are other important exogenous factors in the development of COPD [7,99].

The role of infections in the development of COPD is not well defined [100]. An experimental mouse model of COPD showed that this disease is driven by CD1d-dependent T-cell receptor-invariant natural killer T (NKT) cells, suggesting that the transition from respiratory viral infection into chronic lung disease requires persistent activation of the NKT cell-macrophage innate immune axis [101]. The "AAT Genetic Modifier Study," a multicenter family-based cohort study designed to study the genetic and epidemiological determinants of COPD in 378 adult individuals with AATD, found that cigarette smoking, male sex, asthma, chronic bronchitis, and pneumonia were risk factors for reduced FEV1 and severe airflow obstruction. Remarkably, more than half of the participants reported a history of physician-diagnosed pneumonia [102].

Finally, it is possible that the disease pathogenesis may involve other modifying genes which interacting with environmental factors increase the susceptibility of patients to develop the disease. In this sense, it have been reported several polymorphisms of a number of candidate genes, including that of the nitric oxide synthase (NOS3) [103], glutathione S-transferase (GSTP1) [104], the inhibitor of plasminogen activator SERPINE2 [105] or the antiinflammatory interleukin IL-10 gene [106], but it is not sufficiently proved that any of them play a significant role in the disease pathogenesis.

REFERENCES

[1] Decramer M, Janssens W, Miravitlles M. Chronic obstructive pulmonary disease. Lancet 2012;379:1341–51.

[2] Barnes PJ. Chronic obstructive pulmonary disease. N Engl J Med 2000;343:269–80.

[3] Cosio MG, Saetta M, Agusti A. Immunologic aspects of chronic obstructive pulmonary disease. N Engl J Med 2009;360:2445–54.

[4] Sethi S. Bacterial infection and the pathogenesis of COPD. Chest 2000;117(5 Suppl. 1): 286S–91S.

[5] McCloskey SC, Patel BD, Hinchliffe SJ, Reid ED, Wareham NJ, Lomas DA. Siblings of patients with severe chronic obstructive pulmonary disease have a significant risk of airflow obstruction. Am J Respir Crit Care Med 2001;164:1419–24.

[6] Lieberman J, Winter B, Sastre A. Alpha 1-antitrypsin Pi-types in 965 COPD patients. Chest 1986;89:370e3.

[7] American Thoracic S, European Respiratory S. American Thoracic Society/European Respiratory Society statement: standards for the diagnosis and management of individuals with alpha-1 antitrypsin deficiency. Am J Respir Crit Care Med 2003;168:818–900.

[8] Turino GM, Barker AF, Brantly ML, Cohen AB, Connelly RP, Crystal RG, et al. Clinical features of individuals with PI*SZ phenotype of alpha 1-antitrypsin deficiency. alpha 1-Antitrypsin Deficiency Registry Study Group. Am J Respir Crit Care Med 1996;154(6 Pt 1):1718–25.

[9] Prins J, van der Meijden BB, Kraaijenhagen RJ, Wielders JPM. Inherited chronic obstructive pulmonary disease: new selective-sequencing workup for 1-antitrypsin deficiency identifies 2 previously unidentified null alleles. Clin Chem 2008;54:101–7.

[10] Satoh K, Nukiwa T, Brantly M, Garver RI Jr, Hofker M, Courtney M, et al. Emphysema associated with complete absence of alpha 1-antitrypsin in serum and the homozygous inheritance [corrected] of a stop codon in an alpha 1-antitrypsin-coding exon. Am J Hum Genet 1988;42:77–83.

[11] Fraizer GC, Siewertsen M, Harrold TR, Cox DW. Deletion/frameshift mutation in the alpha 1-antitrypsin null allele, PI*QObolton. Hum Genet 1989;83:377–82.

[12] Lee JH, Brantly M. Molecular mechanisms of alpha1-antitrypsin null alleles. Respir Med 2000;(94 Suppl. C):S7–S11.

[13] Ferrarotti I, Carroll TP, Ottaviani S, Fra AM, O Brien G, Molloy K, et al. Identification and characterisation of eight novel SERPINA1 null mutations. Orphanet J Rare Dis 2014;9:172.

[14] Zorzetto M, Ferrarotti I, Campo I, Balestrino A, Nava S, Gorrini M, et al. Identification of a novel alpha1-antitrypsin null variant (Q0Cairo). Diagn Mol Pathol 2005;14:121–4.

[15] Brantly M, Lee JH, Hildesheim J, Uhm CS, Prakash UB, Staats BA, et al. Alpha1-antitrypsin gene mutation hot spot associated with the formation of a retained and degraded null variant. Am J Respir Cell Mol Biol 1997;16:225–31.

[16] Graham A, Kalsheker NA, Bamforth FJ, Newton CR, Markham AF. Molecular characterisation of two alpha-1-antitrypsin deficiency variants: proteinase inhibitor (Pi) Null(Newport) (Gly115→Ser) and (Pi) Z Wrexham (Ser-19→Leu). Hum Genet 1990;85:537–40.

[17] Nukiwa T, Takahashi H, Brantly M, Courtney M, Crystal RG. alpha 1-Antitrypsin nullGranite Falls, a nonexpressing alpha 1-antitrypsin gene associated with a frameshift to stop mutation in a coding exon. J Biol Chem 1987;262:11999–2004.

[18] Sifers RN, Brashears-Macatee S, Kidd VJ, Muensch H, Woo SL. A frameshift mutation results in a truncated alpha 1-antitrypsin that is retained within the rough endoplasmic reticulum. J Biol Chem 1988;263:7330–5.

[19] Takahashi H, Crystal RG. Alpha 1-antitrypsin Null (isola di procida): an alpha 1-antitrypsin deficiency allele caused by deletion of all alpha 1-antitrypsin coding exons. Am J Hum Genet 1990;47:403–13.

[20] Milger K, Holdt LM, Teupser D, Huber RM, Behr J, Kneidinger N. Identification of a novel SERPINA-1 mutation causing alpha-1 antitrypsin deficiency in a patient with severe bronchiectasis and pulmonary embolism. Int J Chron Obstruct Pulmon Dis 2015;10:891–7.

[21] Faber J-P, Poller W, Weidinger S, Kirchgesser M, Schwaab R, Bidlingmaier F, et al. Identification and DNA sequence analysis of 15 new alpha-1 antitrypsin variants, including 2 PiQO alleles and one deficient PiM allele. Am J Hum Genet 1994;55:1113–21.

[22] Frazier GC, Siewertsen MA, Hofker MH, Brubacher MG, Cox DW. A null deficiency allele of alpha 1-antitrypsin, QOludwigshafen, with altered tertiary structure. J Clin Invest 1990;86:1878–84.

[23] Lara B, Martínez MT, Blanco I, Hernández-Moro C, Velasco EA, Ferrarotti I, et al. Severe alpha-1 antitrypsin deficiency in composite heterozygotes inheriting a new splicing mutation QOMadrid. Respir Res 2014;15:125.

[24] Curiel D, Brantly M, Curiel E, Stier L, Crystal RG. Alpha-1-antitrypsin deficiency caused by the alpha-1-antitrypsin nullmattawa gene. J Clin Invest 1989;83:1144–52.

[25] Lara B, Martínez-Delgado B, Torres ML, Marín-Arguedas S, Bustamante A, Miravitlles M. Alpha-1-antitrypsin deficiency associated with the Mattawa variant. Arch Bronconeumol 2013;49:548–50.

[26] Seixas S, Mendonça C, Costa F, Rocha J. alpha1-Antitrypsin null alleles: evidence for the recurrence of the L353fsX376 mutation and a novel G-->A transition in position +1 of intron IC affecting normal mRNA splicing. Clin Genet 2002;62:175–80.

[27] Vaz Rodrigues L, Costa F, Marques P, Mendonça C, Rocha J, Seixas S. Severe α-1 antitrypsin deficiency caused by Q0(Ourém) allele: clinical features, haplotype characterization and history. Clin Genet 2012;81:462–9.

[28] Hernández Pérez JM, Ramos Díaz R, Fumero García S, Pérez Pérez JA. Description of alpha-1-antitrypsin deficiency associated with PI*Q0ourém allele in La Palma Island (Spain) and a genotyping assay for its detection. Arch Bronconeumol 2015;51:e1–3.

[29] Rametta R, Nebbia G, Dongiovanni P, Farallo M, Fargion S, Valenti L. A novel alpha1-antitrypsin null variant (PiQ0Milano). World J Hepatol 2013;5:458–61.

[30] Brantly M, Hildeshiem J, Laubach V, Rundquist B, Paul L. Alpha-1-antitrypsin nullnew hope: natural conversion of the alpha-1-antitrypsin deficiency Z variant into a null variant by the addition of second amino acid substitution. Clin Res 1992;40:A328.

[31] Brantly M. Alpha-1-antitrypsin genotypes and phenotypes. In: Crystal RG, editor. Alpha-1-antitrypsin deficiency: biology, pathogenesis, clinical manifestations and therapy, vol. 88. New York: Marcel Dekker, Inc.; 1996.

[32] Poller W, Faber JP, Weidinger S, Olek K. DNA polymorphisms associated with a new alpha 1-antitrypsin PIQ0 variant (PIQ0riedenburg). Hum Genet 1991;86:522–4.

[33] Lee J, Novoradovskaya N, Rundquist B, Rewine J, Saltini C, Brantly M. Alpha-1-antitrypsin nonsense mutation associated with retained truncated protein and reduced mRNA. Mol Genet Metab 1998;63:270–80.

[34] Laubach VE, Ryan WJ, Brantly M. Characterization of a human alpha 1-antitrypsin null allele involving aberrant mRNA splicing. Hum Mol Genet 1993;2:1001–5.

[35] Gross P, Pfitzer EA, Tolker E, Babyak MA, Kaschak M. Experimental emphysema: its production with papain in normal and silicotic rats. Arch Environ Health 1965;11:50–8.

[36] Janoff A, Scherer J. Mediators of inflammation in leukocyte lysosomes. IX. Elastinolytic activity in granules of human polymorphonuclear leukocytes. J Exp Med 1968;128:1137–55.

[37] Turino GM, Senior RM, Garg BD, Keller S, Levi MM, Mandl I. Serum elastase inhibitor deficiency and alpha 1-antitrypsin deficiency in patients with obstructive emphysema. Science 1969;165:709–11.

[38] Janoff A. Inhibition of human granulocyte elastase by serum alpha-1-antitrypsin. Am Rev Respir Dis 1972;105:121–2.

[39] Snider GL, Hayes JA, Franzblau C, Kagan HM, Stone PS, Korthy AL. Relationship between elastolytic activity and experimental emphysema-induced properties of papain preparations. Am Rev Respir Dis 1974;110:254–62.

[40] Turino GM, Hornebeck W, Robert B. In vivo effects of pancreatic elastase. I. Studies on the serum inhibitors. Proc Soc Exp Biol Med 1974;146:712–7.

[41] Chrzanowski PJ, Turino GM. Experimental emphysema: concepts and questions. Bull Eur Physiopathol Respir 1977;13:471–7.

[42] Senior RM, Tegner H, Kuhn C, Ohlsson K, Starcher BC, Pierce JA. The induction of pulmonary emphysema with human leukocyte elastase. Am Rev Respir Dis 1977;116:469–75.

[43] Sandhaus RA, Janoff A. Elastase-induced emphysema: retention of instilled proteinase in the rat. Am Rev Respir Dis 1982;126:914–20.

[44] Janoff A, Sloan B, Weinbaum G, Damiano V, Sandhaus RA, Elias J, et al. Experimental emphysema induced with purified human neutrophil elastase: tissue localization of the instilled protease. Am Rev Respir Dis 1977;115:461–78.

[45] Janoff A, Carp H. Possible mechanisms of emphysema in smokers: cigarette smoke condensate suppresses protease inhibition in vitro. Am Rev Respir Dis 1977;116:65–72.

[46] Blue ML, Janoff A. Possible mechanisms of emphysema in cigarette smokers. Release of elastase from human polymorphonuclear leukocytes by cigarette smoke condensate in vitro. Am Rev Respir Dis 1978;117:317–25.

[47] Janoff A, Carp H, Lee DK, Drew RT. Cigarette smoke inhalation decreases alpha 1-antitrypsin activity in rat lung. Science 1979;206:1313–4.

[48] Janoff A, Carp H, Lee DK. Inactivation of alpha 1-proteinase inhibitor and bronchial mucous proteinase inhibitor by cigarette smoke in vitro and in vivo. Bull Eur Physiopathol Respir 1980;(16 Suppl):321–40.

[49] Turino GM. The origins of a concept: the protease-antiprotease imbalance hypothesis. Chest 2002;122:1058–60.

[50] Sandhaus RA, Turino G. Neutrophil elastase-mediated lung disease. COPD 2013;(10 Suppl. 1): 60–3.

[51] Lucey EC, Stone PJ, Breuer R, Christensen TG, Calore JD, Catanese A, et al. Effect of combined human neutrophil cathepsin G and elastase on induction of secretory cell metaplasia and emphysema in hamsters, with in vitro observations on elastolysis by these enzymes. Am Rev Respir Dis 1985;132:362–6.

[52] Kao RC, Wehner NG, Skubitz KM, Gray BH, Hoidal JR. Proteinase 3. A distinct human polymorphonuclear leukocyte proteinase that produces emphysema in hamsters. J Clin Invest 1988;82:1963–73.

[53] Lesser M, Padilla ML, Cardozo C. Induction of emphysema in hamsters by intratracheal instillation of cathepsin B. Am Rev Respir Dis 1992;145:661–8.

[54] Sinden NJ, Baker MJ, Smith DJ, Kreft JU, Dafforn TR, Stockley RA. Alpha-1-antitrypsin variants and the proteinase/anti-proteinase imbalance in chronic obstructive pulmonary disease. Am J Physiol Lung Cell Mol Physiol 2015;308:L179–90.

[55] Burnett D, Abrahamson M, Devalia JL, Sapsford RJ, Davies RJ, Buttle DJ. Synthesis and secretion of procathepsin B and cystatin C by human bronchial epithelial cells in vitro: modulation of cathepsin B activity by neutrophil elastase. Arch Biochem Biophys 1995;317:305–10.

[56] Okada Y, Nakanishi I. Activation of matrix metalloproteinase 3 (stromelysin) and matrix metalloproteinase 2 ('gelatinase') by human neutrophil elastase and cathepsin G. FEBS Lett 1989;249:353–6.

[57] Ferry G, Lonchampt M, Pennel L, de Nanteuil G, Canet E, Tucker GC. Activation of MMP-9 by neutrophil elastase in an in vivo model of acute lung injury. FEBS Lett 1997;402:111–5.

[58] Geraghty P, Rogan MP, Greene CM, Boxio RM, Poiriert T, O'Mahony M, et al. Neutrophil elastase up-regulates cathepsin B and matrix metalloprotease-2 expression. J Immunol 2007;178:5871–8.

[59] Geraghty P, Rogan MP, Greene CM, Brantly ML, O'Neill SJ, Taggart CC, et al. Alpha-1-antitrypsin aerosolised augmentation abrogates neutrophil elastase-induced expression of cathepsin B and matrix metalloprotease 2 in vivo and in vitro. Thorax 2008;63:621–6.

[60] Imamura T, Potempa J, Travis J. Activation of the kallikrein-kinin system and release of new kinins through alternative cleavage of kininogens by microbial and human cell proteinases. Biol Chem 2004;385:989–96.

[61] Döring G. The role of neutrophil elastase in chronic inflammation. Am J Respir Crit Care Med 1994;150:S114–7.

[62] Itoh Y, Nagase H. Preferential inactivation of tissue inhibitor of metalloproteinases-1 that is bound to the precursor of matrix metalloproteinase 9 (progelatinase B) by human neutrophil elastase. J Biol Chem 1995;270:16518–21.

[63] Mihara K, Ramachandran R, Renaux B, Saifeddine M, Hollenberg MD. Neutrophil elastase and proteinase-3 trigger G protein-biased signaling through proteinase-activated receptor-1 (PAR1). J Biol Chem 2013;288:32979–90.

[64] Uehara A, Muramoto K, Takada H, Sugawara S. Neutrophil serine proteinases activate human nonepithelial cells to produce inflammatory cytokines through protease-activated receptor 2. J Immunol 2003;170:5690–6.

[65] Döring G, Frank F, Boudier C, Herbert S, Fleischer B, Bellon G. Cleavage of lymphocyte surface antigens CD2, CD4, and CD8 by polymorphonuclear leukocyte elastase and cathepsin G in patients with cystic fibrosis. J Immunol 1995;154:4842–50.

[66] Bergin DA, Reeves EP, Meleady P, Henry M, McElvaney OJ, Carroll TP, et al. α-1 Antitrypsin regulates human neutrophil chemotaxis induced by soluble immune complexes and IL-8. J Clin Invest 2010;120:4236–50.

[67] Hubbard RC, Fells G, Gadek J, Pacholok S, Humes J, Crystal RG. Neutrophil accumulation in the lung in alpha 1-antitrypsin deficiency. Spontaneous release of leukotriene B4 by alveolar macrophages. J Clin Invest 1991;88:891–7.

[68] Woolhouse IS, Bayley DL, Stockley RA. Sputum chemotactic activity in chronic obstructive pulmonary disease: effect of alpha(1)-antitrypsin deficiency and the role of leukotriene B(4) and interleukin 8. Thorax 2002;57:709–14.

[69] O'Dwyer CA, O'Brien ME, Wormald MR, White MM, Banville N, Hurley K, et al. The BLT1 inhibitory function of α-1 antitrypsin augmentation therapy disrupts leukotriene B4 neutrophil signaling. J Immunol 2015;195:3628–41.

[70] Walsh DE, Greene CM, Carroll TP, Taggart CC, Gallagher PM, O'Neill SJ, et al. Interleukin-8 up-regulation by neutrophil elastase is mediated by MyD88/IRAK/TRAF-6 in human bronchial epithelium. J Biol Chem 2001;276:35494–9.

[71] Devaney JM1, Greene CM, Taggart CC, Carroll TP, O'Neill SJ, McElvaney NG. Neutrophil elastase up-regulates interleukin-8 via toll-like receptor 4. FEBS Lett 2003;544: 129–32.

[72] Deslee G, Woods JC, Moore C, Conradi SH, Gierada DS, Atkinson JJ, et al. Oxidative damage to nucleic acids in severe emphysema. Chest 2009;135:965–74.

[73] Rahman I. The role of oxidative stress in the pathogenesis of COPD: implications for therapy. Treat Respir Med 2005;4:175–200.

[74] Wood AM, Simmonds MJ, Bayley DL, Newby PR, Gough SC, Stockley RA. The TNFalpha gene relates to clinical phenotype in alpha-1-antitrypsin deficiency. Respir Res 2008; 9:52.

[75] Bergin DA, Reeves EP, Hurley K, Wolfe R, Jameel R, Fitzgerald S, et al. The circulating proteinase inhibitor α-1 antitrypsin regulates neutrophil degranulation and autoimmunity. Sci Transl Med 2014;6:217ra1.

[76] Rahman I, Gilmour PS, Jimenez LA, MacNee W. Oxidative stress and TNF-alpha induce histone acetylation and NF-kappaB/AP-1 activation in alveolar epithelial cells: potential mechanism in gene transcription in lung inflammation. Mol Cell Biochem 2002;234-235:239–48.

[77] Mackay F, Loetscher H, Stueber D, Gehr G, Lesslauer W. Tumor necrosis factor alpha (TNF-alpha)-induced cell adhesion to human endothelial cells is under dominant control of one TNF receptor type, TNF-R55. J Exp Med 1993;177:1277–86.

[78] Hurley K, Lacey N, O'Dwyer CA, Bergin DA, McElvaney OJ, O'Brien ME, et al. Alpha-1 antitrypsin augmentation therapy corrects accelerated neutrophil apoptosis in deficient individuals. J Immunol 2014;193:3978–91.

[79] Lee SH, Goswami S, Grudo A, Song LZ, Bandi V, Goodnight-White S, et al. Antielastin autoimmunity in tobacco smoking-induced emphysema. Nat Med 2007;13:567–9.

[80] Rinaldi M, Lehouck A, Heulens N, Lavend'homme R, Carlier V, Saint-Remy JM, et al. Antielastin B-cell and T-cell immunity in patients with chronic obstructive pulmonary disease. Thorax 2012;67:694–700.

[81] Feghali-Bostwick CA, Gadgil AS, Otterbein LE, Pilewski JM, Stoner MW, Csizmadia E, et al. Autoantibodies in patients with chronic obstructive pulmonary disease. Am J Respir Crit Care Med 2008;177:156–63.

[82] Taraseviciene-Stewart L, Scerbavicius R, Choe KH, Moore M, Sullivan A, Nicolls MR, et al. An animal model of autoimmune emphysema. Am J Respir Crit Care Med 2005;171:734–42.

[83] Grudo A, Corry DB, Kheradmand F. An autoimmune basis in the disease pathogenesis of COPD/emphysema (130.41). J Immunol 2007;178:S235–6.

[84] Greene CM, Low TB, O'Neill SJ, McElvaney NG. Anti-proline-glycine-proline or antielastin autoantibodies are not evident in chronic inflammatory lung disease. Am J Respir Crit Care Med 2010;181:31–5.

[85] Wood AM, de Pablo P, Buckley CD, Ahmad A, Stockley RA. Smoke exposure as a determinant of autoantibody titre in α_1-antitrypsin deficiency and COPD. Eur Respir J 2011;37:32–8.

[86] Grumelli S, Corry DB, Song LZ, Song L, Green L, Huh J, et al. An immune basis for lung parenchymal destruction in chronic obstructive pulmonary disease and emphysema. PLoS Med 2004;1:e8.

[87] Sullivan AK, Simonian PL, Falta MT, Mitchell JD, Cosgrove GP, Brown KK, et al. Oligoclonal CD4+ T cells in the lungs of patients with severe emphysema. Am J Respir Crit Care Med 2005;172:590–6.

[88] Baraldo S, Turato G, Lunardi F, Bazzan E, Schiavon M, Ferrarotti I, et al. Immune activation in α1-antitrypsin-deficiency emphysema. Beyond the protease-antiprotease paradigm. Am J Respir Crit Care Med 2015;191:402–9.

[89] Stockley RA. Immune modulation by α1-antitrypsin. A nice concept, but does it influence outcome? Am J Respir Crit Care Med 2015;191:363–4.

[90] Parmar JS, Mahadeva R, Reed BJ, Farahi N, Cadwallader KA, Keogan MT, et al. Polymers of alpha(1)-antitrypsin are chemotactic for human neutrophils: a new paradigm for the pathogenesis of emphysema. Am J Respir Cell Mol Biol 2002;26:723–30.

[91] Mulgrew AT, Taggart CC, Lawless MW, Greene CM, Brantly ML, O'Neill SJ, et al. Z alpha1-antitrypsin polymerizes in the lung and acts as a neutrophil chemoattractant. Chest 2004;125:1952–7.

[92] Mahadeva R, Atkinson C, Li Z, Stewart S, Janciauskiene S, Kelley DG, et al. Polymers of Z alpha1-antitrypsin co-localize with neutrophils in emphysematous alveoli and are chemotactic in vivo. Am J Pathol 2005;166:377–86.

[93] Alam S, Li Z, Janciauskiene S, Mahadeva R. Oxidation of Z {alpha}1-antitrypsin by cigarette smoke induces polymerization: a novel mechanism of early-onset emphysema. Am J Respir Cell Mol Biol 2011;45:261–9.

[94] Alam S, Li Z, Atkinson C, Jonigk D, Janciauskiene S, Mahadeva R. Z α1-antitrypsin confers a proinflammatory phenotype that contributes to chronic obstructive pulmonary disease. Am J Respir Crit Care Med 2014;189:909–31.

[95] Blanco I. Oxidant-mediated aggregation of Z α1-antitrypsin in pulmonary epithelial cells amplifies lung inflammation. Am J Respir Crit Care Med 2014;189:877–9.

[96] Pini L, Tiberio L, Venkatesan N, Bezzi M, Corda L, Luisetti M, et al. The role of bronchial epithelial cells in the pathogenesis of COPD in Z-alpha-1 antitrypsin deficiency. Respir Res 2014;15:112.

[97] Larsson C. Natural history and life expectancy in severe alpha1-antitrypsin deficiency, Pi Z. Acta Med Scand 1978;204:345–51.

[98] Tobin MJ, Cook PJ, Hutchison DC. Alpha 1 antitrypsin deficiency: the clinical and physiological features of pulmonary emphysema in subjects homozygous for Pi type Z. A survey by the British Thoracic Association. Br J Dis Chest 1983;77:14–27.

[99] Eriksson S. A 30-year perspective on alpha 1-antitrypsin deficiency. Chest 1996;110: 237S–42S.

[100] Silverman EK. Exacerbations in chronic obstructive pulmonary disease: do they contribute to disease progression? Proc Am Thorac Soc 2007;4:586–90.

[101] Kim EY, Battaile JT, Patel AC, You Y, Agapov E, Grayson MH, et al. Persistent activation of an innate immune response translates respiratory viral infection into chronic lung disease. Nat Med 2008;14:633–40.

[102] Demeo DL, Sandhaus RA, Barker AF, Brantly ML, Eden E, McElvaney NG, et al. Determinants of airflow obstruction in severe alpha-1-antitrypsin deficiency. Thorax 2007;62:806–13.

[103] Novoradovsky A, Brantly ML, Waclawiw MA, Chaudhary PP, Ihara H, Qi L, et al. Endothelial nitric oxide synthase as a potential susceptibility gene in the pathogenesis of emphysema in alpha1-antitrypsin deficiency. Am J Respir Cell Mol Biol 1999;20:441–7.

[104] Rodriguez F, de la Roza C, Jardi R, Schaper M, Vidal R, Miravitlles M. Glutathione S-transferase P1 and lung function in patients with alpha1-antitrypsin deficiency and COPD. Chest 2005;127:1537–43.

[105] DeMeo D, Mariani T, Lange C, Lake S, Litonjua A, Celedón J, et al. The SERPINE2 gene is associated with chronic obstructive pulmonary disease. Proc Am Thorac Soc 2006;3:502.

[106] Demeo DL, Campbell EJ, Barker AF, Brantly ML, Eden E, McElvaney NG, et al. IL10 polymorphisms are associated with airflow obstruction in severe alpha1-antitrypsin deficiency. Am J Respir Cell Mol Biol 2008;38:114–1120.

Chapter 7

Genetic Epidemiology

7.1 INTRODUCTION

Apart from the most common alpha-1 antitrypsin (AAT) deficiency (AATD) alleles S and Z, about other 50 extremely uncommon "rare" and "null" alleles associated with AATD have been characterized. However, in clinical practice 96% of individuals with diseases associated with AATD exhibit Pi*ZZ genotypes, and only 4% combinations of Z, "rare" or "null" alleles [1–4]. For this reason, and because currently there are not sufficient data on frequencies, prevalence, and geographical distribution of these uncommon "rare" and "null" alleles, the present chapter is limited to the epidemiological analysis of the Pi*S and Pi*Z alleles.

Anecdotally, in 1968, Fagerhol and Tenfjhord performed a pioneering study on allelic frequencies of the AAT gene in Oslo, using an aliquot of serum extracted from seamen of various nationalities who underwent medical examinations, mainly aimed to rule out syphilis, before being hired to temporarily work in Norwegian fishing ships and ports [5]. This survey revealed the existence of very high frequencies of the S allele in Portuguese and Spanish seamen, as well as high Z frequencies in Norwegians (Table 7.1). Despite the small size of most samples of this early study, its findings were premonitory, and in the following years other corroborated that the Z allele is much more common in Caucasians from Northern, Western, and Central Europe (and their descendants settled in other continents) while it is very rare or nonexistent in individuals of yellow and black races, as well as in Amerindians [6–10].

In the late 1990s of last century, Hutchison collected the published studies on the subject in European countries [11], and established for the first time some quality criteria required to rationally consider as valid a study on this topic, that is, (1) sample size of at least 250 subjects; (2) sample composed of healthy, unrelated and randomly chosen individuals; (3) Pi types determined by crossed antigen–antibody electrophoresis (XAA) or isoelectric focusing (IEF), because the starch gel electrophoresis (SG) technique, used in some early studies, had some difficulty for detecting the Z bands of the MZ phenotypes, underestimating the true frequency of the allele Z, and PCR was not been used by any researcher until then; and (4) discarding of screening studies in which phenotypes were determined only in samples with AAT concentration below a low cut-off

TABLE 7.1 Allelic Pi*S and Pi*Z Frequencies in Seamen of Various Nationalities Studied in Norway

Source of subjects	Number	Pi*S frequency (in × 1000)	Pi*Z frequency (in × 1000)
Spain (more than 60% Galicians)	378	112	12
Portugal	39	141	—
Central European countries	186	27	—
Norway	2830	23	16
Latin America	26	58	—
Asia	68	—	—
Africa	9	—	—

Source: Data obtained from Fagerhol MK, Tenfjord OW. Serum Pi types in some European, American, Asian and African populations. Acta Pathol Microbiol Scand 1968;72:601–608 [5].

point, as normal phenotypes were omitted with this methodology while deficiency types were overrepresented [12–16]. For example, the study of Sveger in 200,000 newborns used a cut-off of <40% of AAT, and although this tactic was useful to detect all Z and some SZ phenotypes, it lost all MM and MS, most MZ and SS, and a considerable number of SZ.

A step beyond the calculation of gene frequencies was to analyze them with the Hardy–Weinberg equilibrium formula, an approach used by de Serres and collaborators to provide an estimate for the total number of normal (MM) and deficiency AAT phenotypes (MS, SS, MZ, SZ, and ZZ) in over 500 cohorts, and approximately 200,000 subjects from all races and continents worldwide [17–26]. To try to select only the reliable published studies, de Serres et al. used the same criteria proposed by Hutchison, but in order of assessing the accuracy or inaccuracy of results they added a numerical statistical precision factor (PF) criteria of selection (PF is inversely proportional to the coefficient of variation, which measures the dispersion of values with respect to the mean). However, a remarkable weakness of most selected studies was the lack of uniformity in the composition of the samples studied, the majority newborns, and a variety of "healthy unrelated people," which included blood donors, school or college students, soldiers, hospital staff, persons undergoing preemployment or routine medical examinations, active workers, etc. Only a very small number of studies were conducted in randomly selected general population [27–29]. Interestingly, these studies provided significantly higher values when compared to those provided by surveys performed in not randomly selected "healthy unrelated people" from a same country. Consequently, many of the published results on S and Z frequency may be inaccurate, and so existing data should be regarded only as approximate and taken with caution.

7.2 PI*Z FREQUENCY AND PREVALENCE IN EUROPE

To graphically display the geographical distribution of the Pi*S and Pi*Z frequencies, a series of maps created with the software ArcMap for Microsoft Windows, have been utilized. This technology uses a method of multivariate inverse distance weighting (IDW) interpolation, which generates new data from other known, by an inverse logarithm based on the distance between two points. With this method, Pi*S and Pi*Z frequencies are symbolized by 15–20-colored scales, with blues representing the lowest values and red the highest ones. Areas with maximum frequencies are stained in red and yellow, intermediate in clear yellow and green, and low in a range of blues. The dots represent the places where the selected studies were conducted.

As shown in Fig. 7.1, the maximum frequencies of Pi*Z: 20–40 × 1000 (expressed as the total number of Z alleles per 1000 genes of all Pi types) are in southern Scandinavia, Denmark, Baltic Republics, the Netherlands, Belgium, and Ireland. Values slightly smaller than the previous (15–20 × 1000) appear in Finland, southern England, Western France, and the northwestern regions of the Iberian Peninsula (especially, Galicia, Asturias, and northern of Portugal). In general, values decrease gradually from west to east, and in the most remote regions of the north and south of the continent [30].

In correspondence with the above, the highest prevalence of Pi*ZZ genotypes (1 every 2000–3000 subjects) are found in southern Sweden and Norway, Denmark, Baltic Republics, Finland, Netherlands, Belgium, western of France, south of England, Ireland, and in the Iberian Peninsula. In central Europe the prevalence is 1:4500–5000, and gradually decreases toward the east, and in the most remote regions of the north and south of the continent, with 1:10,000 in the Balkan Peninsula and 1:90,000 in the European Russia, until practically disappear in the eastern part of Russia (Fig. 7.2).

7.2.1 Origin and Spread of the Z Mutation

The Z mutation has a relatively recent origin. It is believed that was generated in Caucasians from southern Scandinavia between 66 and 216 generations (one generation = 30 years), that is, about 2000–6500 years ago [31], though other experts have estimated the occurrence of the Z mutation between 4000 and 7000 years ago [10]. It is accepted that from Scandinavia the Z gene spread to other European countries following major population movements, such as migrations, wars, trade, etc. By the great resemblance between the Z frequency distribution and the Viking settlements during the 8th to the 11th centuries AD, some have some have attributed to this people much of the spread of the mutation to other European populations.

Inhabitants of southern Scandinavia and Denmark, in the 8th and 9th centuries the Vikings had an overpopulation which led to their displacement throughout all the Scandinavian Peninsula, the Baltic regions, and British Isles. In the 9th and 10th centuries they arrived by fluvial courses to the Black Sea,

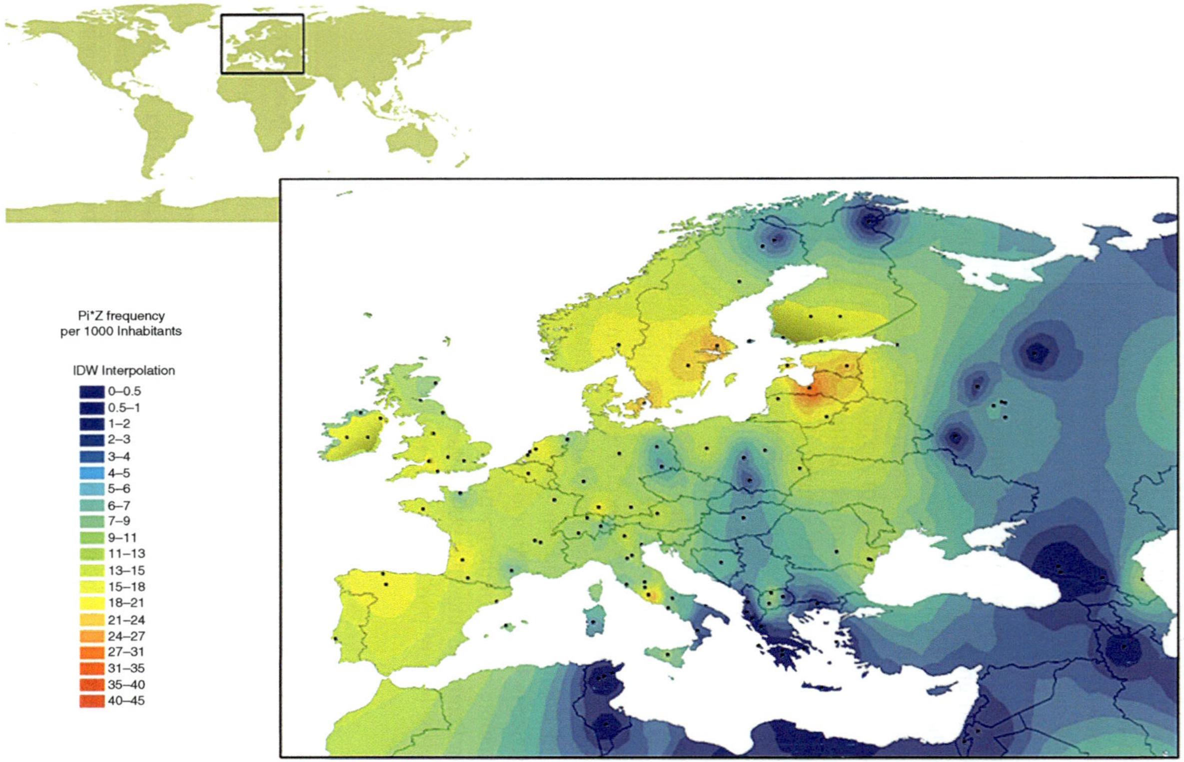

FIGURE 7.1 **Map of Pi*Z allelic frequencies in Europe.** *(Adapted from Blanco I, de Serres FJ, Cárcaba V, Lara B, Fernández-Bustillo E. Alpha-1 antitrypsin deficiency PI*Z and PI*S gene frequency distribution using on maps of the world by an inverse distance weighting (IDW) multivariate interpolation method. Hepat Mon 2012;12(10 HCC):e7434) [30].)*

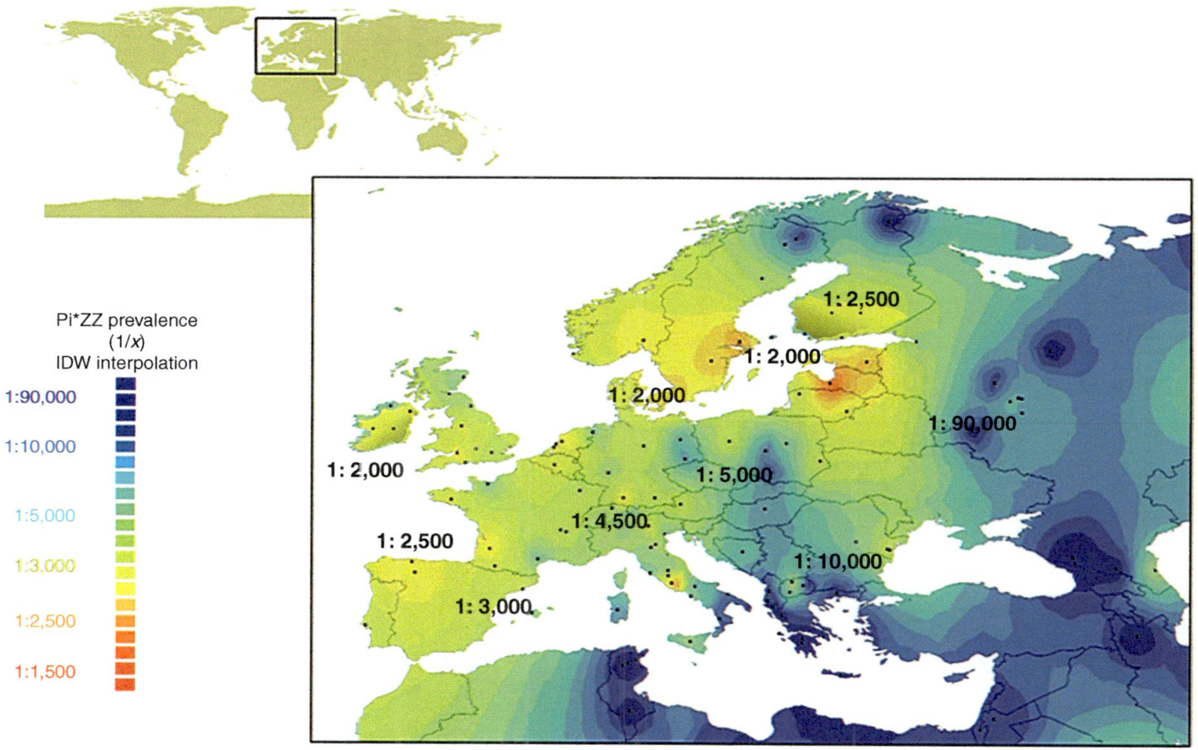

FIGURE 7.2 **Map of Pi*ZZ prevalence in Europe (1:x).** *(Adapted from Blanco I, de Serres FJ, Cárcaba V, Lara B, Fernández-Bustillo E. Alpha-1 antitrypsin deficiency PI*Z and PI*S gene frequency distribution using on maps of the world by an inverse distance weighting (IDW) multivariate interpolation method. Hepat Mon 2012;12(10 HCC):e7434) [30].)*

established a kingdom in Normandy, colonized Iceland, settled in west Greenland, and came to Canada founding a settlement there. In the 11th century, they consolidated the Baltic domain, continued their raids in the Netherlands, the French Atlantic coast, arriving by river to Paris; they devastated the coasts of Galicia and Asturias, Lisbon, Seville, and Mediterranean islands. They also settled in Calabria and Sicily, where they founded some kingdoms. From the 11th century, the Vikings integrated with the populations of the occupied places and ceased their expansion. In the following centuries, thousands of European immigrants carried the gene to the countries where they migrated and settled, especially American countries, Caribbean islands, Australia, and New Zealand.

7.3 PI*Z FREQUENCY IN THE WORLD

On the IDW interpolation map of the World (Fig. 7.3), it can be seen that the greatest number of studies (indicated by points) are in Europe. On the contrary, points are scarce and scattered in the rest of the world, and this lack of studies makes the results less reliable in the areas with limited data.

The maximum frequencies of Pi*Z are in the coastal regions of Europe bathed by the Atlantic Ocean and its seas. They gradually decrease to the east of the continent, virtually disappearing in Asia, with the exception of Pakistan and Thailand, which show moderate values.

In some regions of the northwest, west, and Horn of Africa, there are also moderate values of Pi*Z. In the rest of Africa Z frequencies are absent or very low.

There are moderate frequencies in southeastern Australia. Z frequency is very high in New Zealand.

In North America, there are moderate frequencies in eastern Canada.

In the United States, there are moderate frequencies in the northeastern states and in the states surrounding the Great Lakes and in Arizona. The frequencies are low in Mexico, Central America, Caribbean, and South America [30].

7.4 NUMBER AND DISTRIBUTION OF PI*ZZ GENOTYPES IN EUROPE

In round numbers, it can be summarized that in Europe there may be at least 120,000 Pi*ZZ genotypes, most of them in Germany, France, the Iberian Peninsula, the United Kingdom, Italy, Poland, Belgium, Netherlands, Poland, Ireland, and Northern Europe countries (Fig. 7.4).

7.5 NUMBER AND DISTRIBUTION OF PI*ZZ GENOTYPES IN AMERICA

There could be over 90,000 individuals carrying Pi*ZZ genotypes in North America, Central America, South America, and Caribbean islands (Fig. 7.5).

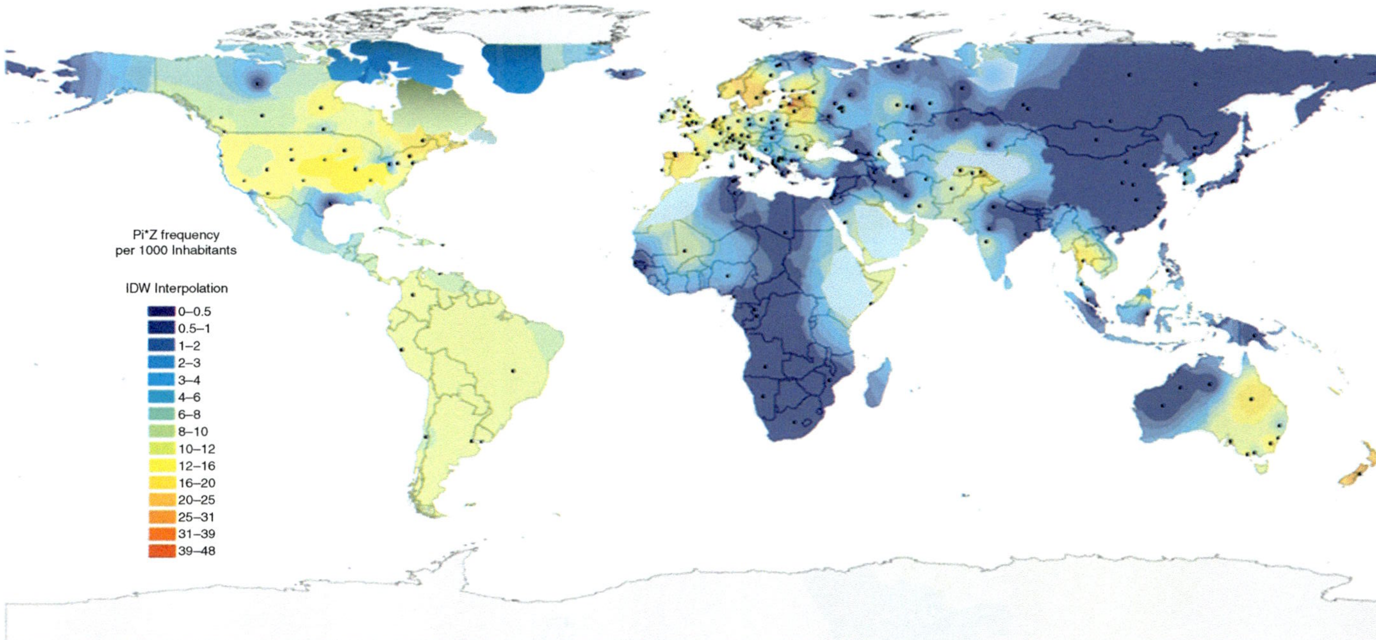

FIGURE 7.3 **Map of Pi*Z allelic frequencies in the world.** *(Adapted from Blanco I, de Serres FJ, Cárcaba V, Lara B, Fernández-Bustillo E. Alpha-1 antitrypsin deficiency PI*Z and PI*S gene frequency distribution using on maps of the world by an inverse distance weighting (IDW) multivariate interpolation method. Hepat Mon 2012;12(10 HCC):e7434) [30].)*

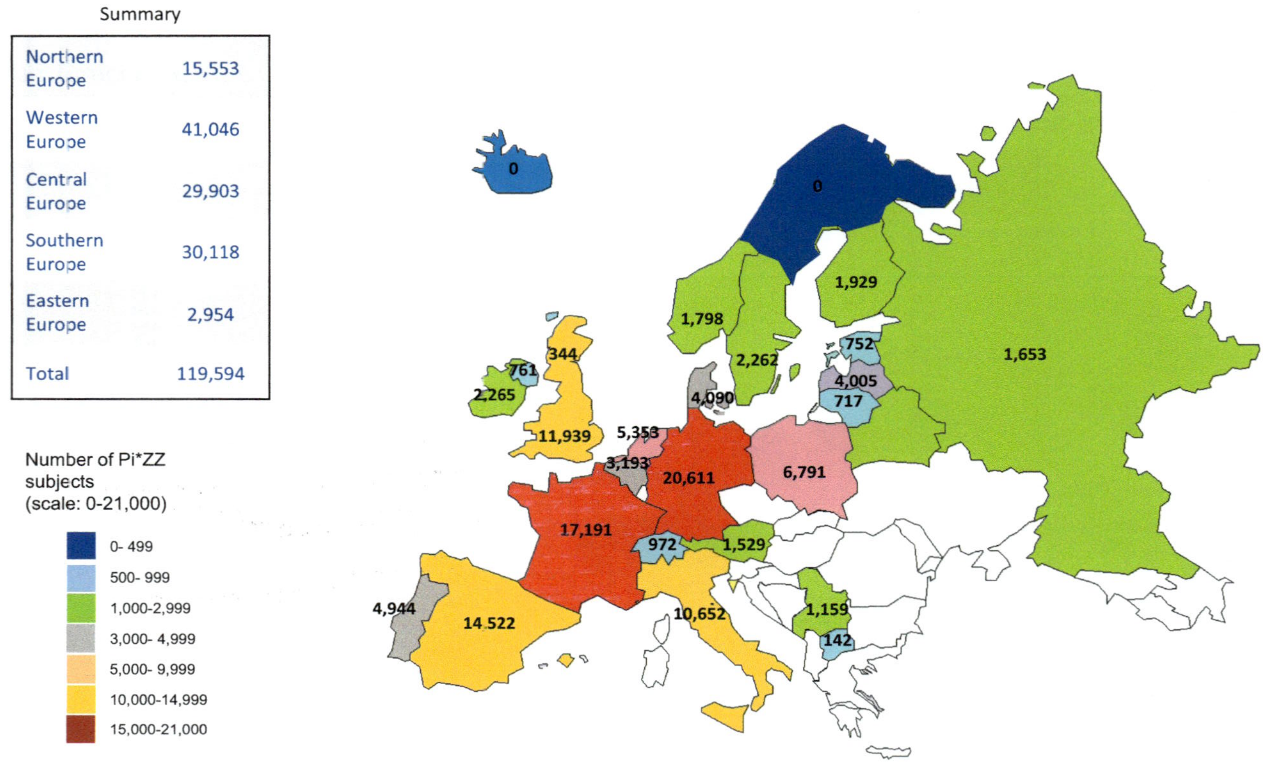

FIGURE 7.4 Approximate number and distribution of genotypes Pi*ZZ in Europe. *(Adapted from de Serres F, Blanco I. Role of alpha-1 antitrypsin in human health and disease. J Intern Med. 2014;276:311–35 [32].)*

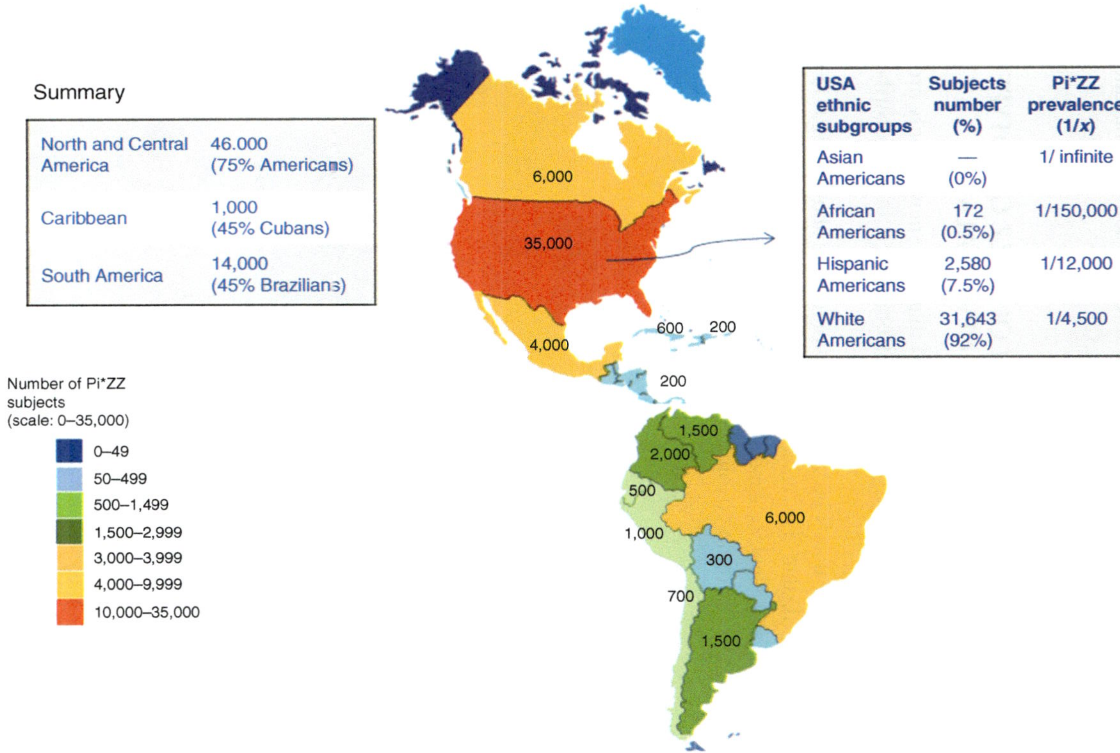

FIGURE 7.5 Estimated number and distribution of Pi*ZZ genotypes in Europe. *(Adapted from de Serres F, Blanco I. Role of alpha-1 antitrypsin in human health and disease. J Intern Med. 2014;276:311–35 [32].)*

Most of these ZZ individuals are in the United States (approximately 70,000, 95% Caucasians).

In addition, it has been estimated that there may be (in round numbers) 7000 in Canada, 4000 in Mexico, 6000 in Brazil, around of 2000 in Colombia Venezuela and Argentina, 1000 in Peru, 700 in Chile, 600 in Cuba, and lesser numbers in other American countries [26,30,32].

7.6 NUMBER AND DISTRIBUTION OF PI*ZZ GENOTYPES IN AFRICA, ASIA, AUSTRALIA, AND NEW ZEALAND

As detailed in Fig. 7.6, it has been estimated that there may be only around of 4000 Pi*ZZ in more than 1.2 billion people from Africa, mainly distributed by Nigeria and Somalia.

In Asia, with more than 4 billion people would have about of 30,000 Pi*ZZ, most of them in Pakistan, Thailand, and South Korea.

In Australia and New Zealand would have about of 6000 Pi*ZZ, most of them descendants of Anglo Saxon settlers [26,30,32].

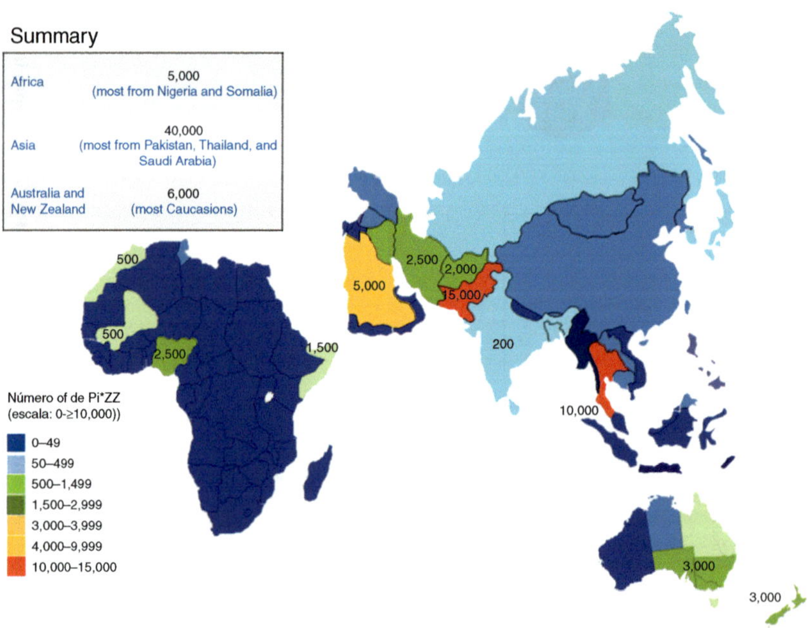

FIGURE 7.6 Estimated number and distribution of Pi*ZZ genotypes in Africa, Asia, Australia, and New Zealand. *(Adapted from de Serres F, Blanco I. Role of alpha-1 antitrypsin in human health and disease. J Intern Med. 2014;276:311–35 [32].)*

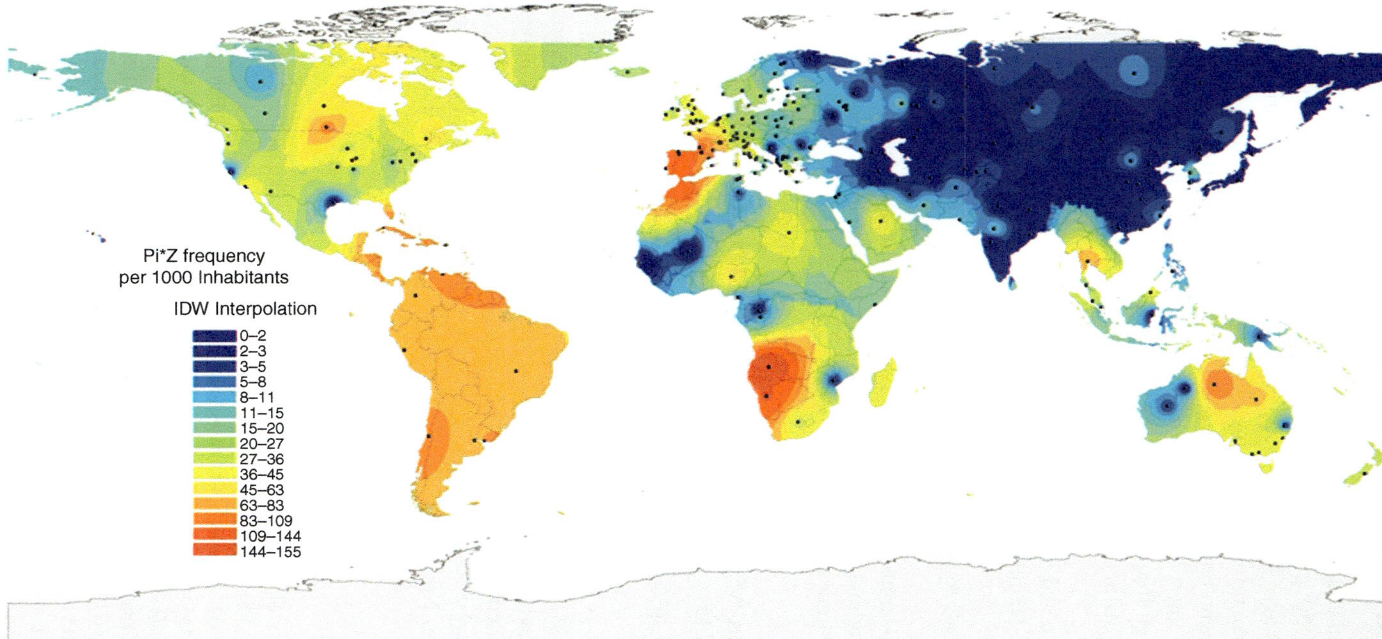

FIGURE 7.7 **Map of Pi*S allelic frequencies in the world.** *(Adapted from Blanco I, de Serres FJ, Cárcaba V, Lara B, Fernández-Bustillo E. Alpha-1 antitrypsin deficiency PI*Z and PI*S gene frequency distribution using on maps of the world by an inverse distance weighting (IDW) multivariate interpolation method. Hepat Mon 2012;12(10 HCC):e7434) [30].)*

7.7 PI*S FREQUENCY DISTRIBUTION IN THE WORLD

It has been postulated that the S allele may have been originated in the Iberian Peninsula 300–450 generations ago (one generation = 30 years), that is, 10–15,000 years ago. It has been suggested that the S variant has arisen in the Franco-Cantabrian refuge area during the last glacial era, spreading from there with the people who repopulated Europe 15,000 years ago [10].

On the IDW interpolation map of the world (Fig. 7.7), it can be seen that there are very high frequencies of Pi*S in the Iberian Peninsula and southwestern France. S frequencies decrease gradually in the north and east of the European continent.

It should be noted very high frequencies in various African regions, especially in Angola, Namibia, Botswana, and South Africa (in which there are not Z alleles).

Except in Saudi Arabia and Thailand, the S allele is virtually nonexistent in the rest of Asia.

There are moderate frequencies of Pi*S in Australia and New Zealand, and in general throughout the American continent populated by descendants of Spanish and Portuguese [26,30,32].

7.8 NUMBER AND DISTRIBUTION OF PI*SZ GENOTYPES IN THE WORLD

The S allele itself carries no significant risk except when combined with Z or with the very uncommon rare and null deficiency alleles. According to the known data it is estimated to be a total of 1,300,000 Pi*SZ genotypes in the countries where AATD has been studied. Since at least 10% of these express serum AAT below the protection value of 11 μM, and therefore they can be at risk of developing diseases associated with AATD.

Approximately 50% of SZ are in Europe (especially in Spain, France, and Portugal), 20% in North America (mostly in USA), 15% in Central and South America, and the remaining 15% in Africa, Asia, and Oceania.

REFERENCES

[1] Lee JH, Brantly M. Molecular mechanisms of alpha1-antitrypsin null alleles. Respir Med 2000;(94 Suppl.):S7–S11.

[2] American Thoracic Society; European Respiratory Society. Thoracic Society/European Respiratory Society Statement: standards for the diagnosis and management of individuals with alpha1-antitrypsin deficiency. Am J Respir Crit Care Med 2003;168:818–900.

[3] Salahuddin P. Genetic variants of alpha1-antitrypsin. Curr Protein Pept Sci 2010;11:101–17.

[4] Ferrarotti I, Carroll TP, Ottaviani S, Fra AM, O'Brien G, Molloy K, et al. Identification and characterisation of eight novel SERPINA1 Null mutations. Orphanet J Rare Dis 2014;9:172.

[5] Fagerhol MK, Tenfjord OW. Serum Pi types in some European, American, Asian and African populations. Acta Pathol Microbiol Scand 1968;72:601–8.

[6] Miravitlles M. Alpha1-antitrypsin deficiency: epidemiology and prevalence. Alpha1-antitrypsin deficiency: epidemiology and prevalence. Respir Med 2000;(94 Suppl. C):S12–5.

[7] Blanco I, Fernández E, Bustillo EF. Alpha-1-antitrypsin PI phenotypes S and Z in Europe: an analysis of the published surveys. Clin. Genet. 2001;60:31–41.

[8] Blanco I, Bustillo EF, Rodríguez MC. Distribution of a1-antitrypsin PI S and PI Z frequencies in countries outside Europe: a meta-analysis. Clin. Genet. 2001;60:431–44.

[9] de Serres FJ. Worldwide racial and ethnic distribution of alpha1-antitrypsin deficiency: summary of an analysis of published genetic epidemiologic surveys. Chest 2002;122:1818–29.

[10] Luisetti M, Seersholm N. Alpha1-antitrypsin deficiency. 1: epidemiology of alpha1-antitrypsin deficiency. Thorax 2004;59:164–9.

[11] Hutchison DC. Alpha 1-antitrypsin deficiency in Europe: geographical distribution of Pi types S and Z. Respir Med 1998;92:367–77.

[12] Sveger T. Liver disease in alpha1-antitrypsin deficiency detected by screening of 200,000 infants. N Engl J Med 1976;294:1316–21.

[13] O'Brien ML, Buist NR, Murphey WH. Neonatal screening for alpha1-antitrypsin deficiency. J Pediatr 1978;92:1006–10.

[14] Dijkman JH, Penders TJ, Kramps JA, Sonderkamp HJ, van den Broek WG, ter Haar BG. Epidemiology of alpha 1-antitrypsin deficiency in the Netherlands. Hum Genet 1980;53:409–13.

[15] Silverman EK, Miletich JP, Pierce JA, Sherman LA, Endicott SK, Broze GJ Jr, et al. Alpha-1-antitrypsin deficiency. High prevalence in the St. Louis area determined by direct population screening. Am Rev Respir Dis 1989;140:961–6.

[16] Alvarez-Granda L1, Cabero-Perez MJ, Bustamante-Ruiz A, Gonzalez-Lamuño D, Delgado-Rodriguez M, García-Fuentes M. PI SZ phenotype in chronic obstructive pulmonary disease. Thorax 1997;52:659 61.

[17] de Serres FJ, Blanco I, Fernández-Bustillo E. Genetic epidemiology of alpha-1 antitrypsin deficiency in southern Europe: France, Italy, Portugal and Spain. Clin Genet 2003;63:490–509.

[18] de Serres FJ, Blanco I, Fernández-Bustillo E. Genetic epidemiology of alpha-1 antitrypsin deficiency in North America and Australia/New Zealand: Australia, Canada, New Zealand and the United States of America. Clin Genet 2003;64:382–97.

[19] de Serres FJ, Blanco I, Fernández-Bustillo E. Health implications of alpha1-antitrypsin deficiency in Sub-Sahara African countries and their emigrants in Europe and the New World. Genet Med 2005;7:175–84.

[20] Ferrarotti I, Blanco I, Fernández-Bustillo E. Alpha-1 antitrypsin deficiency in Italy: regional differences of the PIS and PIZ deficiency alleles. de Serres FJ, Luisetti M. Monaldi Arch Chest Dis 2005;63:133–41.

[21] Blanco I, de Serres FJ, Fernandez-Bustillo E, Lara B, Miravitlles M. Estimated numbers and prevalence of PI*S and PI*Z alleles of alpha1-antitrypsin deficiency in European countries. Eur Respir J 2006;27:77–84.

[22] de Serres FJ, Blanco I, Fernández-Bustillo E. Estimated numbers and prevalence of PI*S and PI*Z deficiency alleles of alpha1-antitrypsin deficiency in Asia. Eur Respir J 2006;28:1091–9.

[23] de Serres FJ, Blanco I, Fernández-Bustillo E. PI S and PI Z alpha-1 antitrypsin deficiency worldwide. A review of existing genetic epidemiological data. Monaldi Arch Chest Dis 2007;67:184–208.

[24] de Serres FJ, Blanco I, Fernández-Bustillo E. Estimates of PI*S and PI*Z Alpha-1 antitrypsin deficiency alleles prevalence in the Caribbean and North, Central and South America. Monaldi Arch Chest Dis 2009;71:96–105.

[25] de Serres FJ, Blanco I, Fernández-Bustillo E. Ethnic differences in alpha-1 antitrypsin deficiency in the United States of America. Ther Adv Respir Dis 2010;4:63–70.

[26] de Serres FJ, Blanco I. Prevalence of α1-antitrypsin deficiency alleles PI*S and PI*Z worldwide and effective screening for each of the five phenotypic classes PI*MS, PI*MZ, PI*SS, PI*SZ, and PI*ZZ: a comprehensive review. Ther Adv Respir Dis 2012;6:277–95.

[27] Blanco I, Fernández E, Rodríguez MC, Fernández A. [Allelic frequency of the gene of alpha-1-antitrypsin in the general population in a county in Asturias]. Med Clin (Barc) 1999;113:366–70.

[28] Carroll TP, O'Connor CA, Floyd O, McPartlin J, Kelleher DP, O'Brien G, et al. The prevalence of alpha-1 antitrypsin deficiency in Ireland. Respir Res 2011;12:91.

[29] Häggblom J, Kettunen K, Karjalainen J, Heliövaara M, Jousilahti P, Saarelainen S. Prevalence of PI*Z and PI*S alleles of alpha-1-antitrypsin deficiency in Finland. Eur Clin Respir J 2015;2:28829.

[30] Blanco I, de Serres FJ, Cárcaba V, Lara B, Fernández-Bustillo E. Alpha-1 antitrypsin deficiency PI*Z and PI*S gene frequency distribution using on maps of the world by an inverse distance weighting (IDW) multivariate interpolation method. Hepat Mon 2012;12(10 HCC):e7434.

[31] Cox DW, Woo SL, Mansfield T. DNA restriction fragments associated with alpha 1-antitrypsin indicate a single origin for deficiency allele PI Z. Nature 1985;316:79–81.

[32] de Serres F, Blanco I. Role of alpha-1 antitrypsin in human health and disease. J Intern Med 2014;276:311–35.

Chapter 8

Laboratory Diagnosis

8.1 INTRODUCTION

In general, the clinical use of electrophoresis in plasma protein analysis is based on the electrophoretic separation of plasma proteins, according to their mobility, into albumin, alpha1 (α1), alpha2 (α2), beta (β), and gamma (γ) globulins, in spite that each of these electrophoretic zones contains two or more major plasma proteins (e.g., α1: α1-antitrypsin (90%) and α1-acid glycoprotein; α2: haptoglobin, α2-macroglobulin, α2-antiplasmin, and ceruloplasmin; β: transferrin, low-density lipoprotein and complement; and γ: IgA, IgD, IgE, IgG, and IgM immunoglobulins) [1].

The first laboratory diagnosis of alpha1-antitrypsin (AAT) deficiency (AATD) was performed by Laurell and Eriksson in 1963 using a combination of agarose gel and paper electrophoresis [2] (Fig. 8.1). The agarose gel electrophoresis had important limitations, including the difficulty to characterize some heterozygous phenotypes, such as MZ, MS, or SZ, and was soon abandoned for this reason. Shortly afterward, Fagerhol developed a method of starch gel electrophoresis at acid pH, but as the previous electrophoretic method it was no longer routinely used for its difficulty to identify MZ heterozygotes [3]. Subsequently, the agarose and starch gel techniques were replaced by isoelectric focusing (IEF) of serum in polyacrylamide gels, which provided much better resolution [4], even more enhanced since the development of immobilized pH gradients, in the 1980s [5,6]. In the next decade, the molecular analysis of the gene (i.e., genotyping) was progressively introduced as a reference method for identifying rare and null allelic variants, for the characterization of new variants and to study some cases where there was no correlation between AAT levels and the phenotype characterized by IEF [7,8].

Dried blood spot (DBS) specimens on filter paper allow quantification of AAT by immunonephelometry, identification of the AAT phenotype by IEF, and AAT genotyping, and are now being successfully used for large-scale AATD diagnostics at reasonable cost. In addition, its use as a screening method is increasingly becoming popular among pneumologists [9–12].

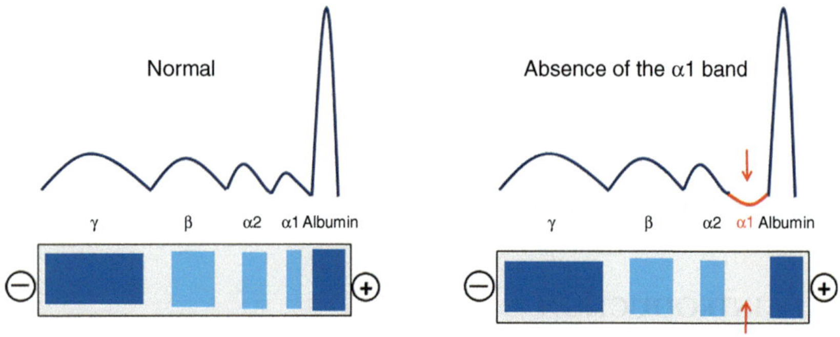

FIGURE 8.1 Schematic representation of a normal and an alpha-1 antitrypsin deficiency electrophoresis gel. The absence of alpha-1 band is quite evident when comparing the normal electrophoretic pattern (left) with the alpha-1 antitrypsin deficiency pattern (right).

8.2 CLINICAL SITUATIONS IN WHICH ALPHA-1 ANTITRYPSIN DEFICIENCY SHOULD BE SUSPECTED

Diagnosis of AATD is based primarily on a high index of suspicion on the part of clinicians who treat these patients. In general, clinical features that suggest the possibility of AATD include COPD (usually emphysema) at an early age, especially in patients with the absence of a recognized risk factor like smoking or occupational dust exposure, severe bronchial asthma with persistent airflow obstruction, unexplained liver disease in all ages of life, cytoplasmic antineutrophil cytoplasmic antibody-positive vasculitis, and neutrophilic panniculitis. The existence of other familial cases with these same diseases, especially if they have already been diagnosed with AATD, strongly increases the decision of testing the suspects. Once the disease is suspected, laboratory evaluation is mandatory, and involves quantitation of circulating AAT and characterization of the AAT polymorphism (i.e., phenotyping or genotyping) (Table 8.1).

The observation of a reduced or absent alpha-globulin band on the routine plasma protein electrophoresis should increase suspicion of AATD, which must be confirmed with quantitative and qualitative specific test, that is, measurement of AAT serum level, and IEF or DBS real-time PCR and melting curves, depending on availability of the laboratory.

Remarkably, scientific societies recommend to measure serum levels of AAT at least once in life to all COPD subjects [8,13,14].

8.3 MEASUREMENT OF SERUM ALPHA1-ANTITRYPSIN LEVELS

The first step to reach the diagnosis is to perform a blood (plasma or serum) determination of the AAT level, a simple, cheap, and widely available test. AAT level can be determined by rocket immunoelectrophoresis, radial immunodiffusion, or nephelometry. Currently, the analysis most commonly performed is a nephelometry in fresh or frozen serum samples, avoiding repeated freezing and thawing [8].

TABLE 8.1 **Clinical Situations in Which Alpha-1 Antitrypsin Deficiency Should be Suspected**

Lung diseases

• Scientific societies recommend to measure serum levels of AAT at least once in life to all COPD subjects, especially if:
  • *Early onset emphysema*
  • *No smokers*
  • *Family history of COPD*
  • *Family history of AAT deficiency*
• Severe bronchial asthma of difficult control

Liver diseases
• Infants and children with chronic liver disease of unknown etiology • Adults with cirrhosis or chronic hepatitis of unknown etiology   • *Especially if there is a family history of AAT deficiency*
Other diseases
• Systemic vasculitis   • *Especially if antineutrophil cytoplasmic antibodies positive (ANCA+)* • Neutrophilic panniculitis

AAT, Alpha-1 antitrypsin; COPD, chronic obstructive pulmonary disease.

Radial immunodiffusion tends to overestimate the AAT concentration by as much as 40%, and though less, nephelometry may also overestimate AAT levels in case of interference with lipids or hemoglobin. The turbidity or some samples after repeated thawing may also give falsely high values [8].

To discriminate between historic values obtained using the nonpurified standard and those obtained with the pure standard developed by the US Registry, the former are expressed in milligrams per deciliter (mg/dL) and the latter in micromolar units (μmol/L or μM) [15] (Table 8.2). The reference range of the serum AAT concentration for adults is usually between 100 and 220 mg/dL, although it may differ slightly between one laboratory to another, and some reference laboratories (especially from the United States) can report serum levels in micromolar units, with a normal range of 20–60 μmol/L. Nevertheless, these values are interchangeable; to convert μM in mg/dL, the micromolar concentration should be multiplied by a conversion factor of 5.2, and to convert a value obtained by nephelometry into micromolar units it should be multiplied by 0.1923. Anyway, for a good interpretation, each laboratory must ideally have previously established the normal values of AAT in serum samples obtained from a healthy general population individuals [16].

To interpret the result of an isolated quantitative determination, it must be take into account that the AAT is an acute phase reactant, so that infectious or inflammatory processes can alter the determination raising levels and giving falsely normal or high values in patients with moderate deficit. High levels of serum AAT during pregnancy and after consumption of oral contraceptives have also been reported [8].

TABLE 8.2 Plasma Concentration Ranges of the Alpha-1 Antitrypsin Expressed by Different Protease Inhibitor (Pi) Genotypes

Pi genotype	Purified standard US Registry (μM/L)	Nephelometry (mg/dL)	Radial immunodiffusion (mg/dL)
MM	20–38	103–200	150–350
MS	19–35	100–180	120–250
SS	13–20	70–105	100–200
MZ	13–23	66–120	75–210
SZ	9–15	45–80	65–120
ZZ	2–8	10–40	20–45
Z-Rare; Rare–Rare	2–8	10–40	20–45
Z-Null; Rare–Null	2–3	10–15	15–35
Null–Null	≈0	≈0	≈0

Values are expressed as micromolar units (μmol/L or μM) for concentrations measured using the purified standard from the US Registry and in milligrams per deciliter (mg/dL) for measurements performed by commercial standards determined by nephelometry or radial immunodiffusion techniques. To convert μM in mg/dL, the micromolar value must be multiplied by a conversion factor of 5.2, and to convert mg/dL into micromolar units the mg/dL values (determined by nephelometry) must be multiplied by 0.1923.

Serum concentrations of AAT for children are slightly lower than those of adults, but the differences are not likely clinically relevant [17].

It should be noted that a "protective" threshold level of 11 μmol/L corresponds to 80 mg/dL if measured by radial immunodiffusion and to 57.2 mg/dL if measured by nephelometry (the conversion factor to convert micromolar to milligrams per deciliter is the AAT molecular weight: 5.2; i.e., 11 × 5.2 = 57.2). Alternatively, to convert 11 μmol/L from the International System of Units (SI) to conventional units (mg/dL), the value in micromoles is divided by the conversion factor 0.184; that is, 11/0.184=59.8 mg/dL (http://www.globalrph.com/conv_si.htm). Therefore, for orientation purposes the 11 μmol/L equivalence for mg/dL may be between 50 and 60, but possibly closer to the latter value. This protective threshold has evolved from the observation that patients with heterozygote phenotypes whose levels of AAT exceed this level are usually free from emphysema, although the scientific bases supporting this assumption are observational and devoid of sufficient statistical power [18].

Recently, it has been described a new method, yet at experimental level, which can differentiate between S and Z and M proteins, allowing for quantitation of circulating wild-type AAT in heterozygous carriers of the S and Z deficiency alleles [19]. In this preliminary study, allele-specific quantitation showed significant differences of wild-type protein expression in MZ and MS patients, and thus its potential future application may be important, for

Protein	Liver polymerization (%)	Serum concentration (%)	K (M/s)
M	0	100	1.4×10^7
S	55	45	1.1×10^7
Z	85	15	7.3×10^6

K: constant (coefficient) of chemical reaction rate AAT against neutrophil elastase, expressed in per mole per liter per second

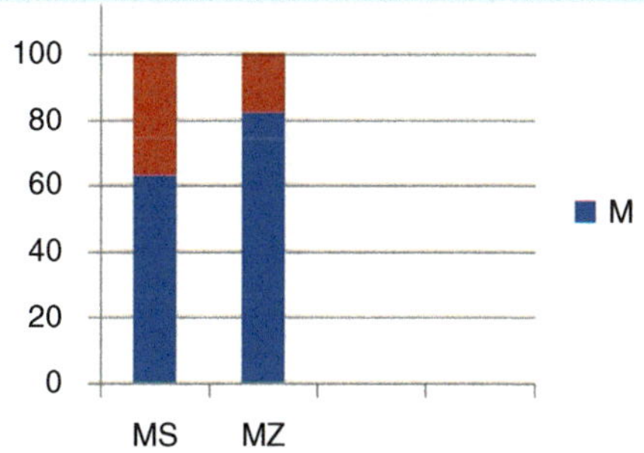

FIGURE 8.2 Liver polymerization, serum concentration, chemical reaction rate (K) of M, S, and Z proteins, and proportion of these proteins in the total AAT of MS and MZ heterozygotes. Compared with the normal M and the deficient S, Z protein has important disadvantages: (1) increased polymerization, (2) reduction of serum and tissue concentration, and (3) decreased antiprotease activity. The bar chart shows that the total serum alpha-1 antitrypsin of a group of MS subjects was composed by 63% of M and 37% of S protein, while the alpha-1 antitrypsin of MZ subjects had 82% of M and 18% of Z protein. *(Data taken from Donato LJ, Karras RM, Katzmann JA, Murray DL, Snyder MR. Quantitation of circulating wild-type alpha-1-antitrypsin in heterozygous carriers of the S and Z deficiency alleles. Respir Res 2015;16:96 [19].)*

example, in SZ subjects with low expression of AAT, because protein Z has several disadvantages. The first one is a profuse polymerization with 85% retention into the producing cells (particularly hepatocytes) with a secondary proportional decrease of serum and tissues concentration. In addition, the serious conformational changes experienced by the Z molecule result in an 80% of functional activity loss [20,21] In contrast, the S protein polymerizes much less, maintains a serum and tissue levels of approximately 45%, and retains its inhibitory capacity [22] (Fig. 8.2).

8.4 IDENTIFICATION OF ALPHA-1 ANTITRYPSIN PHENOTYPES

Measurement of AAT levels can identify patients with protein deficiency, but cannot differentiate between the various AATD genetic subtypes. The most widely used method for identifying AAT variants is IEF [23].

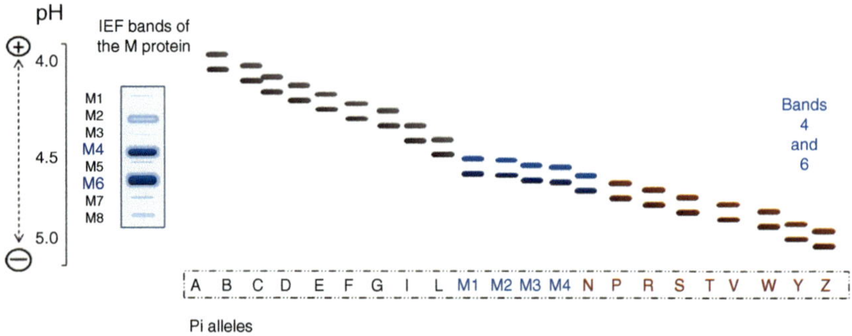

FIGURE 8.3 Schematic representation of the alpha-1 antitrypsin major bands (4 and 6) in various alleles from the protease inhibitor (Pi) system. By exposing the alpha-1 antitrypsin to the effect of a magnetic field with a pH range between 4 and 5, the M protein breaks down into 8 bands (in this example M1–8). In the scheme, the bands 4 and 6 of the anodic variants (A–L) are colored in *gray*, the 4 and 6 cathodic bands form variants N–Z in *brown*, and the M major bands (in the center of the magnetic field) in *blue*.

Phenotyping may be performed on serum or plasma samples. This technique involves the electrophoretic separation of proteins according to their isoelectric point in an agarose gel and in a 4.2–4.9 pH gradient [8]. M variant migrates to the central part of the magnetic field; anodic variants are designated by the letters of the alphabet from A to L; and the cathodic variants with the letters N–Z (Fig. 8.3).

By exposing the AAT to the effect of a magnetic field with a fixed acid pH range of approximately 4–5, the protein breaks down into 8 bands numbered from 1 to 8, according to their isoelectric point (i.e., the pH at which the charge of the molecule becomes zero). These bands are classified into major and minor, the former being more intensely stained. Bands 4 and 6 are the major bands; 2, 7, and 8 the minor ones; and 1, 3, and 5 in general do not appear or are hardly perceptible in IEF. AAT bands are named with the letter of the variant followed by the band number (Fig. 8.4A).

Phenotyping is indicated in all cases with AAT levels below an expected normal range [8] because a specific phenotype from a given AAT concentration cannot safely be inferred, due to fact that among several of them there is a marked overlapping of values (Table 8.2). In any case it is highly advisable that each laboratory has their own reference values of AAT concentrations for each of the major phenotypes.

Phenotyping is the current gold standard for detecting Pi*S and Pi*Z variants, but unfortunately, this technique is time-consuming, requires considerable skill and experience, and is usually done only in well-equipped laboratories limiting its availability to a few reference services. Storage conditions of serum samples are very important as they influence the result, particularly for identifying PiZ phenotypes, whose antigens are very labile and tend to degrade [8].

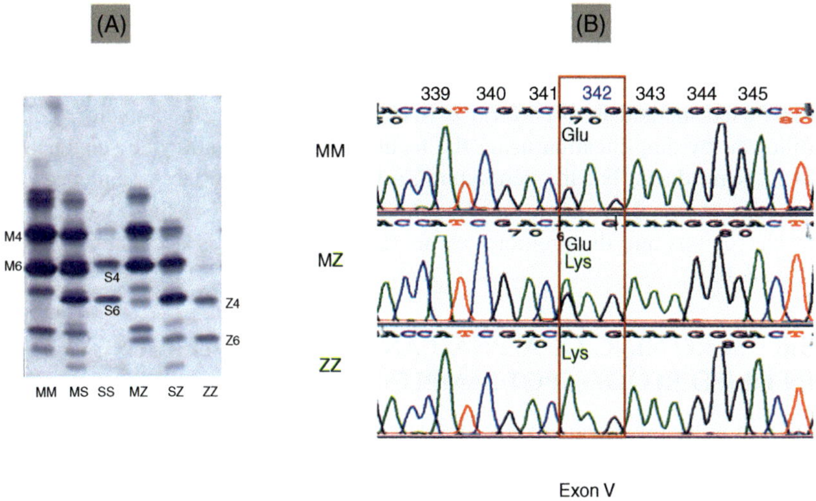

FIGURE 8.4 Isoelectric focusing patterns of the alpha-1 antitrypsin MM, MS, SS, MZ, SZ, and ZZ phenotypes, and fragment of the alpha-1 antitrypsin exon V sequence, showing the normal MM and the MZ and ZZ deficiency variants. (A) The M4 and M6 major bands of the normal MM phenotype migrate to the middle of the magnetic field. The S4 and S6 bands of the Pi*S phenotype, and the Z4 and Z6 of the Pi*Z type, migrate to the cathodic area of the magnetic field, and therefore they are represented at different levels below the M bands. (B) Fragment of the of exon V sequence of the alpha-1-antitrypsin gene. The upper panel corresponds to the normal variant MM, the central to the moderately deficiency MZ, and the lower to the severe deficiency ZZ. In the Z mutation there is a substitution of a base in an AAG codon by GAG (adenine by glycine), resulting in transcription of lysine instead of glutamic acid in position 342.

A limitation of the IEF is that some rare variants, such as Pi*Mmalton, Pi*Mpalermo, Pi*Mduarte, and so on, have an isoelectric point similar to the normal allele Pi*M, and for this reason they cannot be catalogued by phenotyping. Moreover, IEF technique cannot detect null alleles, does not identify new variants, and cannot differentiate between a Pi*ZZ and a Pi*Z-Null phenotype. For these reasons, in cases where there is no correlation between AAT levels and phenotyping, a molecular characterization of the gene by genotyping techniques should be performed [24].

Some laboratories perform IEF on "dried blot spot" samples, using a blood drop absorbed on special paper, allowing for easier transport of samples. This method is suitable for screening purposes, but the identification of a deficient variant should be confirmed on serum or plasma samples [8–12].

8.5 ALPHA-1 ANTITRYPSIN GENOTYPING

Molecular analysis of the AAT gene is the reference method for identifying rare allelic variants associated with congenital deficiency of AAT, for characterization of new and null variants, and for the study of cases in which there is no correlation between the AAT concentration and the IEF phenotype [8].

The determination is usually carried out from whole blood, using EDTA as anticoagulant. Samples can be stored at 4°C for 48 h after performing extraction or stored at −20°C or −80°C if they are not processed immediately. The most used method is to perform a DNA extraction from mononuclear cells, followed by amplification using the technique of PCR and cycle sequencing of PCR products. For the genotyping, a full study of the DNA sequences of the four exons of the gene encoding AAT (II–V), as well as the proximal intronic regions and the promoter exons (IA, IB and IC), is necessary [25,26] (Fig. 8.4B).

8.6 SCREENING OF ALPHA-1 ANTITRYPSIN DEFICIENCY IN DRIED BLOOD SPOT SAMPLES

DBS samples are especially useful for screening and diagnosis of genetic diseases. Obtaining venous blood by finger stick (or finger prick) is minimally invasive, being easy DBS storage and shipment to a reference laboratory where the determinations are centralized. The blood drops are applied on paper discs that are left to dry at room temperature. In some protocols, the concentration of AAT is currently determined by a semiquantitative method of kinetic nephelometry, and only diagnoses the two most frequent deficiency alleles S and Z [9–12].

The detection is performed using allele-specific PCR in real-time and specific "fluorescence resonance energy transfer" probes, or Taqman hydrolysis probes, or restriction pattern genotyping techniques [27–30]. The allele-specific PCR real-time techniques are fast, easily automatable, and low cost. However, they only report the presence or absence of the studied alleles (usually S and Z), but their results cannot exclude the presence of other deficiency alleles. Therefore, it is recommended that the results of the AATD screening-diagnosed cases be corroborated by quantitative determination of AAT and phenotyping in serum samples, and if necessary by genotyping in total blood or DBS samples [31] (Fig. 8.5).

Recently, genotyping using PCR methodologies is moving forward and can now detect S, Z, Mmalton, and F variants, and a next generation of diagnostic kits that can measure most AATD mutations may be available in the near future [26,32].

8.7 NEONATAL SCREENING AND PRENATAL DIAGNOSIS

Given the potential benefits and the risks of its application, the ATS/ERS guidelines do not recommend population screening or predispositional fetal testing [8].

Prenatal diagnosis will require amniotic fluid obtained by amniocentesis or chorionic villus sampling to provide material on which genetic testing be performed. In addition, prenatal diagnosis would require techniques of DNA

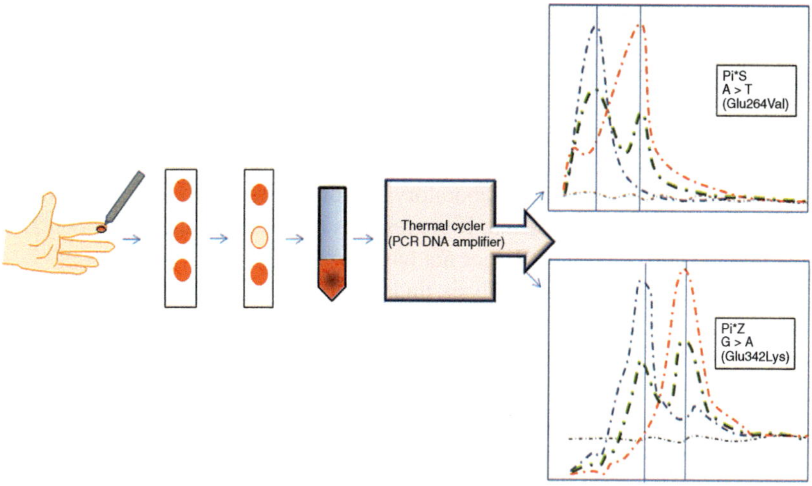

FIGURE 8.5 Molecular diagnostics of Pi*S and Pi*Z variants by real-time PCR and melting curves in dried blood spot samples fixed on paper. Dried blood spot (DBS) samples obtained by puncture of the fingertip of a hand finger are used for allele-specific PCR in real-time and specific "fluorescence resonance energy transfer" (FRET) probes, or Taqman hydrolysis probes, or restriction pattern genotyping techniques. These techniques report the presence or absence of the alleles studied. In the figure, some examples of "melting peaks" obtained by real-time PCR and FRET probes for detection of the S and Z variants have been drawn.

amplification and specific probes to provide adequately specific diagnostics, and to our knowledge it is beyond the scope of the vast majority laboratories [8].

8.8 DETERMINING THE ABILITY OF THE ALPHA-1 ANTITRYPSIN TO INHIBIT NEUTROPHIL ELASTASE

The elastase inhibitory activity is related to the concentration of AAT, so that the larger amount of circulating AAT the greater inhibitory effect. However, some AATD patients may have normal levels of AAT, but little antielastase activity. In these extremely rare circumstances, specialized laboratories can perform a functional assay which measures the ability of the patient's serum to inhibit human leukocyte elastase. This test is used to evaluate a patient with clinical features that are highly suggestive of AATD but whose serum levels are within the reference range [8].

8.9 LABORATORY LIVER FUNCTION TESTS

The initial evaluation of liver function in patients with AATD is performed by determinations of serum transaminases, bilirubin, albumin, and coagulation tests (activated partial thromboplastin time and international normalized ratio). If any of these analyses were altered, the patient would have to complete a broader study at the discretion of a specialist in hepatology.

8.10 CONCLUSIONS

AATD is diagnosed using a combination of genetic and biochemical tests.

The key to reach a diagnosis of AATD is that the practicing physician suspects that his patient's disease may be related to the AATD, requesting a measurement of the AAT blood concentration to his laboratory.

Once the sample is in the laboratory, the first thing to do is to measure the AAT levels in serum or plasma. If the result is normal, the study ends concluding that the tested individual does not have AATD.

On the contrary, if the analysis finds AAT levels below the normal range (e.g., <100 mg/dL, although the reference value may vary between the laboratories), it should be analyzed by qualitative IEF electrophoresis (phenotyping). Using this standard diagnostic procedure, the great majority of AATD phenotypes are identified. Exceptionally, if serum levels are not concordant with the reported phenotype, it should be necessary to make a whole blood gene sequencing (genotyping) to identify rare or null genotypes.

In the event that a physician might not have the aforementioned resources, DBS samples are suitable for screening purposes, but the identification of a deficient variant by this alternative diagnostic procedure should be confirmed with serum or plasma samples, following the standard procedure (Fig. 8.6).

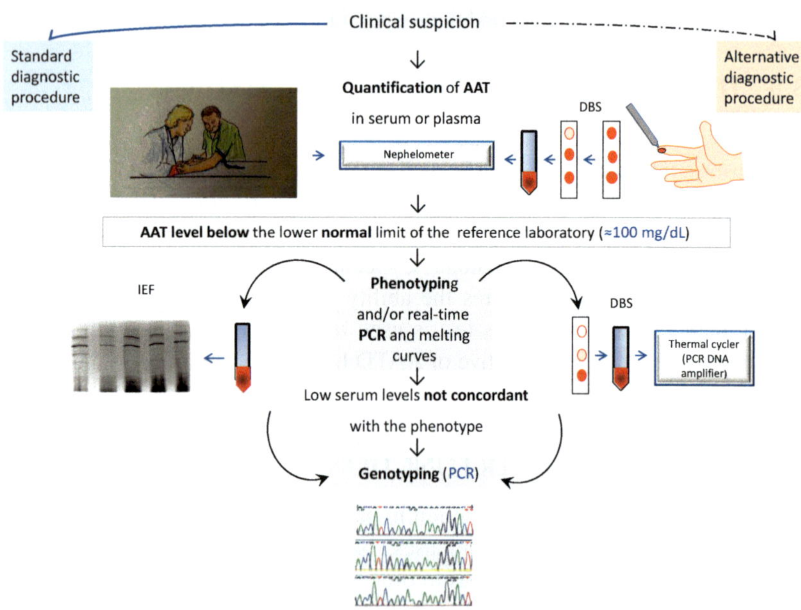

FIGURE 8.6 Standard and alternative steps to diagnose a patient with alpha-1 antitrypsin deficiency.

REFERENCES

[1] Jacoby RF, Cole CE. Molecular diagnostic methods in cancer genetics. In: Abeloff MD, editor. Clinical oncology. 2d ed. New York: Churchill Livingstone; 2000. p. 119–21.

[2] Laurell CB, Eriksson S. The electrophoretic alpha1-globulin pattern of serum in alpha1-antitrypsin deficiency. Scand J Clin Lab Invest 1963;15:132–40.

[3] Fagerhol MK. The Pi-system, genetic variants of serum 1 antitrypsin. Series Haematol 1968;151:153–61.

[4] Allen RC, Harley RA, Talamo RC. A new method for determination of alpha-1-antitrypsin phenotypes using isoelectric focusing on polyacrylamide gel slabs. Am J Clin Pathol 1974;62:732–9.

[5] Görg A, Postel W, Weser J, Patutschnick W, Cleve H. Improved resolution of PI (alpha 1-antitrypsin) phenotypes by a large-scale immobilized pH gradient. Am J Hum Genet 1985;37:922–30.

[6] Righetti PG, Gianazza E, Bianchi-Bosisio A, Sinha P, Köttgen E. Isoelectric focusing in immobilized pH gradients: applications in clinical chemistry and forensic analysis. J Chromatogr 1991;569:197–228.

[7] Okayama H, Holmes MD, Brantly ML, Crystal RG. Characterization of the coding sequence of the normal M4 alpha 1-antitrypsin gene. Biochem Biophys Res Commun 1989;162:1560–70.

[8] American Thoracic Society/European Respiratory Society. American Thoracic Society/European Respiratory Society statement: standards for the diagnosis and management of individuals with alpha-1 antitrypsin deficiency. Am J Respir Crit Care Med 2003;168:818–900.

[9] Aslanidis C, Nauck M, Schmitz G. High-speed detection of the two common alpha(1)-antitrypsin deficiency alleles Pi*Z and Pi*S by real-time fluorescence PCR and melting curves. Clin Chem 1999;45:1872–5.

[10] Costa X, Jardi R, Rodriguez F, Miravitlles M, Cotrina M, Gonzalez C, et al. Simple method for alpha1-antitrypsin deficiency screening by use of dried blood spot specimens. Eur Respir J 2000;15:1111–5.

[11] Rodriguez F, Jardí R, Costa X, Cotrina M, Galimany R, Vidal R, et al. Rapid screening for alpha1-antitrypsin deficiency in patients with chronic obstructive pulmonary disease using dried blood specimens. Am J Respir Crit Care Med 2002;166:814–7.

[12] Struniawski R, Szpechcinski A, Poplawska B, Skronski M, Chorostowska-Wynimko J. Rapid DNA extraction protocol for detection of alpha-1 antitrypsin deficiency from dried blood spots by real-time PCR. Adv Exp Med Biol 2013;756:29–37.

[13] WHO [alpha 1-Pi, deficiency]. World Health Organization. Pneumologie 1997;51:885–918.

[14] Casas F, Blanco I, Martínez MT, Bustamante A, Miravitlles M, Cadenas S, et al. Indications for active case searches and intravenous alpha-1 antitrypsin treatment for patients with alpha-1 antitrypsin deficiency chronic pulmonary obstructive disease: an update. Arch Bronconeumol 2015;51:185–92.

[15] Vidal R, Blanco I, Casas F, Jardí R, Miravitlles M. Committee on the National Registry of Individuals with Alpha-1 Antitrypsin Deficiency. [Guidelines for the diagnosis and management of alpha-1 antitrypsin deficiency]. Arch Bronconeumol 2006;42:645–59.

[16] Bornhorst JA, Greene DN, Ashwood ER, Grenache DG. alpha1-Antitrypsin phenotypes and associated serum protein concentrations in a large clinical population. Chest 2013;143:1000–8.

[17] Donato LJ, Jenkins SM, Smith C, Katzmann JA, Snyder MR. Reference and interpretive ranges for α(1)-antitrypsin quantitation by phenotype in adult and pediatric populations. Am J Clin Pathol 2012;138:398–405.

[18] Turino GM, Barker AF, Brantly ML, Cohen AB, Connelly RP, Crystal RG, et al. Clinical features of individuals with PI*SZ phenotype of alpha 1-antitrypsin deficiency. Alpha 1-Antitrypsin Deficiency Registry Study Group. Am J Respir Crit Care Med 1996;154(6 Pt. 1):1718–25.

[19] Donato LJ, Karras RM, Katzmann JA, Murray DL, Snyder MR. Quantitation of circulating wild-type alpha-1-antitrypsin in heterozygous carriers of the S and Z deficiency alleles. Respir Res 2015;16:96.

[20] Ogushi F, Fells GA, Hubbard RC, Straus SD, Crystal RGOgushi F, Hubbard RC, et al. Z-type alpha 1-antitrypsin is less competent than M1-type alpha 1-antitrypsin as an inhibitor of neutrophil elastase. J Clin Invest 1987;80:1366–74.

[21] Sinden NJ, Baker MJ, Smith DJ, Kreft JU, Dafforn TR, Stockley RA. α-1-Antitrypsin variants and the proteinase/antiproteinase imbalance in chronic obstructive pulmonary disease. Am J Physiol Lung Cell Mol Physiol 2015;308:L179–90.

[22] Ogushi F, Hubbard RC, Fells GA, Casolaro MA, Curiel DT, Brantly ML, et al. Evaluation of the S-type of alpha-1-antitrypsin as an in vivo and in vitro inhibitor of neutrophil elastase. Am Rev Respir Dis 1988;137:364–70.

[23] Ferrarotti I, Scabini R, Campo I, Ottaviani S, Zorzetto M, Gorrini M, et al. Laboratory diagnosis of alpha1-antitrypsin deficiency. Transl Res 2007;150:267–74.

[24] Zerimech F, Hennache G, Bellon F, Barouh G, Jacques Lafitte J, Porchet N, et al. Evaluation of a new Sebia isoelectrofocusing kit for alpha 1-antitrypsin phenotyping with the Hydrasys System. Clin Chem Lab Med 2008;46:260–3.

[25] Brantly M. Laboratory diagnosis of alpha-1 antitrypsin deficiency. In: Crystal RG, editor. Alpha 1-antitrypsin deficiency. New York: Marcel Dekker; 1996. p. 211–26.

[26] Miravitlles M, Herr C, Ferrarotti I, Jardi R, Rodriguez-Frias F, Luisetti M, et al. Laboratory testing of individuals with severe alpha1-antitrypsin deficiency in three European centres. Eur Respir J 2010;35:960–8.

[27] Ferrarotti I, Zorzetto M, Scabini R, Mazzola P, Campo I, Luisetti M. A novel method for rapid genotypic identification of alpha 1-antitrypsin variants. Diagn Mol Pathol 2004;13:160–3.

[28] Bartels CL, Marchetti AL, Edward Highsmith W, Tsongalis GJ. Real time PCR detection of the PI*Z and PI*S mutations associated with alpha-1 antitrypsin deficiency. Am J Transl Res 2009;1:406–11.

[29] Balduyck M, Chapuis Cellier C, Roche D, Odou MF, Joly P, Madelain V, Vergne A, Nouadje G, Lafitte JJ, Porchet N, Beaune P, Zerimech F. Development of a laboratory test on dried blood spots for facilitating early diagnosis of alpha-1-antitrypsin deficiency. Ann Biol Clin (Paris) 2014;72:689–704.

[30] Wittwer CT, Ririe KM, Andrew RV, David DA, Gundry RA, Balis UJ. The LightCycler: a microvolume multisample fluorimeter with rapid temperature control. Biotechniques 1997;22:176–81.

[31] Lyon E, Wittwer CT. LightCycler technology in molecular diagnostics. J Mol Diagn 2009;11:93–101.

[32] Bals R, Koczulla R, Kotke V, Andress J, Blackert K, Vogelmeier C. Identification of individuals with alpha-1-antitrypsin deficiency by a targeted screening program. Respir Med 2007;101:1708–14.

Chapter 9

Clinical Diagnosis

9.1 INTRODUCTION

Alpha-1-antitrypsin (AAT) deficiency (AATD) is an inherited condition that confers predisposition to various diseases throughout life, mainly chronic obstructive pulmonary disease (COPD) and various types of liver diseases (including neonatal cholestasis, juvenile hepatitis, liver cirrhosis in children and adults, and hepatocellular carcinoma in adults). Much less frequently, AATD is associated with systemic vasculitis and neutrophilic panniculitis. Some people may experience more than one of these AATD-related diseases, while others do not suffer from any throughout their entire life [1].

There is insufficient evidence on whether the presence of deficiency alleles influences the frequency or severity of bronchial asthma or bronchiectasis, and the lack of scientific evidence is even greater with respect to many other diseases of inflammatory, autoimmune, or tumoral nature, which occasionally have been related to AATD on the basis of isolated observations, small case series, or case–control studies with insufficient statistical significance [1–3] (Table 9.1).

The most typical picture of the AATD is that of a young adult, smoker or not, with rapidly progressive dyspnoea and early emphysema development. Other patients are diagnosed at later ages after years of suffering from nonspecific symptoms of COPD (i.e., cough, sputum production, dyspnoea, and frequent bronchial exacerbations). In smokers, the onset of symptoms may occur in the third or fourth decade of life, whereas in nonsmokers COPD symptoms are delayed until the sixth decade. Remarkably, a small number of nonsmokers (and exceptionally some smokers) never manifest symptoms and have a normal lifespan, being diagnosed by chance in family studies. Some patients have a positive bronchodilator test which raises the differential diagnosis between COPD and bronchial asthma [1].

Liver pathology has a bimodal distribution (Fig. 9.1). Only a small percentage of ZZ homozygous begin with cholestasis, jaundice, and elevated liver enzymes in early childhood. A minority of these can progress to cirrhosis and potentially fatal liver failure if liver transplantation is not attempted. During adolescence, youth, and early adulthood, most patients are free of symptoms. Hepatic involvement is much more common in males after 50 years of age, and their frequency increases gradually with age. Some patients may develop hepatocellular carcinomas, usually at advanced ages both in cirrhotic and noncirrhotic livers [1].

Blanco's Overview of Alpha-1 Antitrypsin Deficiency

TABLE 9.1 Alpha-1 Antitrypsin Deficiency-Related Diseases

Degree of acceptance among experts	Disease and penetrance (%)
Unanimously accepted	• COPD (≈60%) • Liver diseases: children cholestasis (10%); child and adolescent cirrhosis (2.5%); adult cirrhosis (<50 years: 10%; >50 years: 20%–40%); hepatocarcinoma (2%–3%) • Systemic vasculitis c-ANCA+ (2%) • Neutrophilic panniculitis (0.1%–0.9%)
Conflicting opinions for insufficient or contradictory evidence	• Bronchiectasis (27%) • Bronchial asthma (?)
Very doubtful association for lack of significant evidence	• Rheumatoid arthritis, inflammatory bowel disease, fibromyalgia, aneurysms and arterial dissections, psoriasis, chronic urticaria, mesangiocapillary glomerulonephritis, pancreatitis, anterior uveitis, diabetes mellitus, hypothyroidism, multiple sclerosis, cancers (lung, colorectal, gallbladder, lymphomas, and so on), and possibly other inflammatory and autoimmune diseases

c-ANCA, Cytoplasmic antineutrophil cytoplasmic antibody.
The "degree of acceptance by experts" has been deduced from the content of Refs. [1–7].

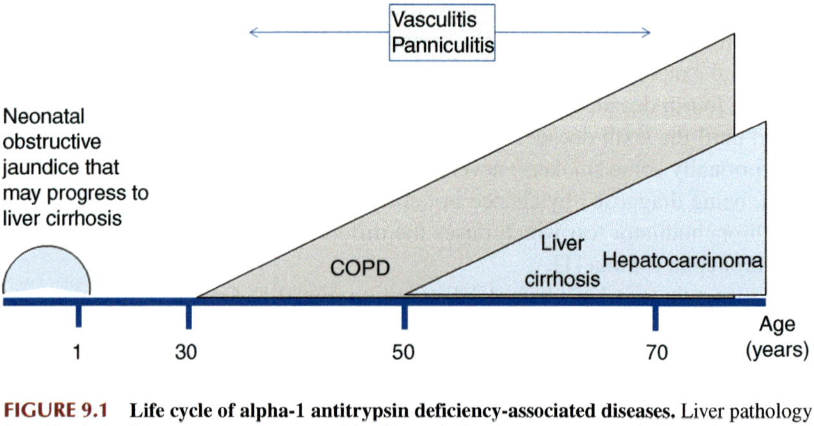

FIGURE 9.1 **Life cycle of alpha-1 antitrypsin deficiency-associated diseases.** Liver pathology (usually cirrhosis) can occur at any age of the life cycle, but is more common in the early years and after 50 years. Pulmonary disease (typically COPD) begins to manifest from 30 years, especially in smokers, and its frequency increases with the passage of the years. Vasculitis and panniculitis are rare manifestations, and can affect both young people and adult persons. *(Figure adapted from the courtesy of Maurizio Luisetti.)*

Between 1 and 9 out of 1000 subjects with severe AATD (usually Caucasians with a Pi*ZZ genotype) develops neutrophilic panniculitis, manifested by painful recurrent ulcerating subcutaneous nodules, histologically characterized by dense infiltrates of neutrophils in the deep dermis and connective tissue septae, with secondary lobular panniculitis. Panniculitis may be the only clinical manifestation of AATD, although it can also occur together with the classical pulmonary or hepatic manifestations of the disease [4].

Finally, between 2% and 3% of patients with severe AATD develops systemic vasculitis with multiple organ involvement [1–3].

9.2 RISK FOR DISEASES ASSOCIATED WITH THE DIFFERENT ALPHA-1 ANTITRYPSIN DEFICIENCY GENOTYPES

The risk for liver and lung disease associated with the most common SERPINA1 deficiency genotypes (compared with the normal MM genotype) is summarized in Table 9.2.

The MS genotype is associated with a normal serum concentration of AAT and no increased risk of liver or lung disease development.

The SS genotype is associated with a slight decrease in the serum concentration of AAT, and does not appear to be associated with an increased risk for diseases, but more appropriate studies than those available are needed to confirm this assumption [1].

The MZ genotype is characterized by a moderate decrease in serum concentration of AAT and is not considered to be at increased risk for diseases, but

TABLE 9.2 Alpha-1 Antitrypsin Pi* Genotypes, Liver Polymers Accumulation, Serum Concentrations, and Risk for Liver Disease and COPD Development

Pi* genotype	Polymers accumulation into the liver	Risk for liver disease	AAT serum level (mg/dL)	Risk for COPD
MM	No	Normal	100–220	Normal
MS	Very low	Normal	100–180	Normal
SS	Low	Normal	70–105	Unknown
MZ	Moderate	No/very low	66–120	Doubtful[a]
SZ	Abundant	Increased[a]	45–80	Increased[a]
ZZ	Very abundant	Very increased	10–40	Very high
Null–Null	No	Normal	—	Extremely high

AAT, Alpha-1 antitrypsin; Pi*, protease inhibitor; COPD, chronic obstructive pulmonary disease.
[a]*If other genetic or environmental risk modified factors are associated with this genotype.*

emerging data suggest that a small subset of individuals with this genotype may suffer from accelerated lung destruction, especially if they are smokers. The results of the available studies on this topic are contradictory, with case–control studies generally finding some increased risk for COPD, and population-based studies finding no differences in lung function between MZ and MM subjects. It is therefore still uncertain whether all MZ individuals are at increased risk for COPD or if only a subset of them are at increased risk due to other environmental or genetic factors of susceptibility [5].

The SZ genotype is associated with a severe decrease in serum concentration of AAT and an increased risk for liver and lung disease in approximately 20% of individuals whose AAT serum levels are below the protective threshold value of 57–60 mg/dL, especially if they are smokers [6,7].

Finally, it is well established that the ZZ genotype is associated with intense intrahepatic polymerization, very severe decrease in serum concentration of AAT and high risk for liver and lung diseases [8].

Apart from the ZZ, other deficiency rare alleles associated with intrahepatic polymerization and inclusion bodies formation (e.g., Mmalton, Siiyama, etc.) are also at increased risk for developing liver and lung diseases [9].

Null homozygotes are not at risk of developing liver disease because they do not produce AAT protein, and consequently do not cause intrahepatic deposits of mutant proteins; however, they are associated with high risk of developing obstructive airways diseases [10,11].

In clinical practice, it is widely accepted that the risk of disease development is mostly linked to the ZZ genotype, although occasionally it can also be linked to SZ, rare, and null deficiency genotypes with serum concentrations of AAT below the "protective threshold value" of 11 μM/L or 60 mg/dL (measured by nephelometry) [1,7]. Although this value has been obtained from an observational study biased by the small sample size and nonrandom selection of the subjects studied [12], a more recent study in 699 Pi*ZZ, 126 Pi*SZ, and 316 Pi*MM subjects from the UK Registry has also found the same value of 11 μM/L for "protective threshold value" of serum AAT [6].

Existing data on penetrance (i.e., percentage of severe AATD individuals that develop clinical illness) are insufficient to draw conclusions. Nevertheless, it is known that practically all null phenotypes develop COPD, and it seems evident that up to 60% of ZZ individuals may develop chronic airflow obstruction, especially if they are smokers. The intriguing variability of penetrance indicates that AATD by itself may not always be enough to facilitate the development of diseases, but rather it must be associated with other genetic and/or environmental favoring factors [1,7–9].

9.3 "ALPHA-1 ANTITRYPSIN DEFICIENCY IS NOT A RARE DISEASE BUT A DISEASE THAT IS RARELY DIAGNOSED"

According to the celebrated sentence of Professor de Serres (1929–2015) that heads this section, AATD is a hereditary condition too often forgotten by clinicians, whose detection is in most cases not performed or performed too late, depriving patients of the benefits derived from an early diagnosis, which may allow to do a special effort in smoking cessation, to identify other at-risk family members, to provide genetic counseling, and to consider augmentation therapy if available and indicated [13].

However, this statement cannot be equally applied to all countries because, according to several recently published data from 10 major European registries and the Canadian Registry, there are very significant differences between countries on the number of cases with severe AATD registered and the number of estimated cases [14–24]. Thus, while in Belgium, France, and Spain the percentage of registered cases is less than 5% of that expected, in Canada, Germany, Ireland, Italy, Poland, and United Kingdom this percentage ranges from 5% to 15%, rises to 20% in the Netherlands, and reaches the remarkable rate of 75% in Sweden, the country in which the AATD was discovered, and where there is much awareness for this condition in both the Swedish population and the public health services (Table 9.3).

Despite the fact that there is a remarkable lack of reliable epidemiological data, the overall estimates of AATD subjects worldwide appear impressive, suggesting that AATD is probably one of the most common hereditary

TABLE 9.3 Number of Registered Subjects With Alpha-1 Deficiency in Different National Registries and Their Percentage to the Estimated Number of Cases

Country	Estimated number of Pi*ZZ subjects	No. of registered subjects with severe AAT deficiency[a]	Percentage of registered cases V.S. total estimated
Belgium	2917	55	2
Canada	5592	290	5
France	9794	273	3
Germany	7839	1066	14
Ireland	2015	204	10
Italy	3481	422	12
The Netherlands	2388	485	20
Poland	660	55	8
Spain	12,125	469	4
Sweden	2077	1553	75
United Kingdom	11,503	930	8

AAT, Alpha-1 antitrypsin.
[a]*Most of registered subjects are Pi*ZZ, and only a minority are severe subjects with rare or null genotypes.*

underrecognized disorders in the world, with only a small minority of subjects currently detected [25].

AATD diagnosis is often delayed in patients with pulmonary disease because their symptoms are identical to those of other more common obstructive lung diseases, such as asthma in younger individuals or COPD in older individuals, a fact which is very misleading and makes the screening for AATD not considered [26].

Underdiagnoses and underreporting of cases are a widespread problem worldwide that has been tried to explain by the following reasons: (1) the low index of suspicion of some health professionals attributable to lack of knowledge or lack of interest in this topic; (2) the belief that diagnostic methods are too complicated and not very affordable for doctors who work outside of referral hospitals; (3) the variability and lack of specificity of the AATD clinical presentation, with between one- and two-thirds of ZZ asymptomatic or misdiagnosed as COPD erroneously solely attributed to smoking, nonresponsive asthma, or cirrhosis only attributed to alcohol consumption without investigating AATD; (4) the lack of a specific treatment in countries where augmentation therapy is not approved, which discourages physicians to screen for the disease; (5) the objections about cost and efficacy of replacement therapy given by some health officials that hamper its use; and (6) the concern of some doctors about possible personal, family, and social negative repercussions (including stigmatization) that could result in discovering a noncurable hereditary disease in a patient [27–29].

As a result of underdiagnoses, the diagnostic average delay between the onset of symptoms and the time of diagnosis varies ordinarily between 6 and 10 or even more years, resulting in an average age at diagnosis of approximately 45 years [30]. Regrettably, although targeted detection of AATD has been advocated by most official society guidelines, and a variety of measures had been undertaken in the last years [including educational campaigns, provision of free test kits, issuance of reminders with medical reports, placing reminders in pulmonary function services, empowering respiratory therapists to conduct testing for AATD in pulmonary function laboratories, and stimulating detection in people with COPD using dried blood spot (DBS) samples in primary care centers], it seems that the situation has not changed significantly [31,32].

The World Health Organisation (WHO) [33] and the American Thoracic Society/European Respiratory Society recommendations [1] state that all symptomatic adults with persistent airway obstruction should be screened for AATD. However, the observed discrepancy between the number of expected patients and the number of identified cases leads to the pessimistic assumption that these recommendations are not being rigorously followed [34].

In summary, available data highlight the need to continue education of medical staff regarding AATD, with the purpose of enhancing awareness, disseminate knowledge, and adopt appropriate measures to implement the existing recommendations.

9.4 GENERAL POPULATION AND NEWBORN SCREENINGS

In the years following the discovery of AATD numerous studies were conducted to detect AATD in healthy people from the general population, as well as in patients with diverse types of diseases (both related and unrelated to this hereditary condition), and in newborns. Their results regarding the detection of cases were generally poor, and the used methodologies expensive and laborious, so the current trend has drastically changed in the sense of directing screening studies to groups of patients with AATD-related diseases, especially COPD or chronic liver disease of unknown etiology. Nevertheless, despite its little usefulness to detect cases, some of these initial studies were useful to provide a rough estimate of the M, S, and Z allelic distribution in many countries of the world [35].

Currently, massive screenings in newborns and in healthy individuals from the general population have become obsolete and only justified in regions with high prevalence of AATD and significant underdiagnosis. To carry them out, it is recommended to do so with the support of scientific societies, expert supervision, and adequate information for volunteers, both before and after the results, and to get an informed consent signed by each participant. In addition, before ordering the genetic test, the physician should weigh the pros (e.g., early diagnosis, preventive measures, genetic counseling, specific replacement therapy) and cons (e.g., psychological trauma, and family, social, and labor negative impact, including possible stigmatization) that could cause the detection of this inherited condition in a healthy person [35].

9.5 TARGETED DETECTION

A targeted high-risk case finding approach is a relatively feasible and cost-effective tool to increase detection of AATD in individuals with obstructive lung diseases (i.e., emphysema, persistent airflow obstruction, and chronic bronchitis), liver disease at any age, granulomatosis with polyangiitis, neutrophilic panniculitis, and family members of patients with confirmed AATD [33].

The joint ATS/ERS statement recommends targeted testing for AATD in all individuals included in any of these groups, especially those with COPD, liver disease of unknown etiology, and relatives of cases with AATD [1,33].

9.6 FAMILY STUDIES

The ATS/ERS guidelines recommend evaluation of first-degree relatives (i.e., siblings, parents, and children) of individuals with severe AATD, to identify as early as possible those who would benefit from therapeutic and preventive measures [1]. Extended pedigree analysis beyond first-degree relatives may be indicated in a more distant family member (beyond parents, sibs, and offspring) suffering from COPD, liver cirrhosis, vasculitis, or panniculitis of unknown cause [7].

9.7 PRENATAL DIAGNOSIS

In general, prenatal diagnosis (or prenatal screening) consists in testing for diseases or conditions in a fetus or embryo before it is born, with the aim of detecting genetic disorders to enable treatment of a condition before or after birth, or to give the parents the chance to abort a fetus with the diagnosed serious and incurable condition, or to get parents ready for a baby with a health problem or for the likelihood of a stillbirth [36]. In the AATD context, prenatal genetic testing may be required in the case of an expectant parents who are carriers of Pi*Z or null mutations. This procedure would be advisable in the case of pregnant homozygous ZZ or with rare mutations, especially to some at-risk couples who have previously had a child with severe liver disease. However, such testing is not useful in predicting if the genetic condition will manifest clinically or not, the age of onset, or the severity of the disorder [7].

If both parents are Pi*MZ, the risk to offspring of having AATD and severe liver disease in childhood ranges from 1% to 10% [7,37].

Fetal testing is not recommended in the American Thoracic Society/European Respiratory Society guidelines because of the variable expressivity of disease and the possibility that some AATD subjects do not express any disease throughout their life [1,7].

The diagnosis is established by studying the genotype by direct analysis of fetal DNA in maternal blood fetal cells [38], transabdominal chorionic villus samples, amniotic fluid obtained by amniocentesis, and other methods of prenatal screening and diagnosis of varying invasiveness [39].

9.8 ALGORITHMS FOR ALPHA-1 ANTITRYPSIN DEFICIENCY DIAGNOSIS

Currently, there is not a rigid sequence of steps unanimously accepted to identify the AATD in the laboratory. Nevertheless, most experts agree that the combined use of different lab procedures, depending on their availability, provides acceptably good results [40–42].

As it has been discussed in the previous chapter of this book (Fig. 8.6), to confirm the diagnosis of AATD in a patient with a clinical suspicion, according to the classical algorithm, a determination of serum (or plasma) protein levels by nephelometry should be performed initially. If AAT serum levels are normal, AATD can be discarded, and further studies are not necessary because the person studied is very probably going to have a normal MM phenotype. However, if serum levels are below the lower normal reference limit, the diagnosis of AATD is confirmed, and next step should be to characterize the AAT-Pi phenotype, by means of isoelectric focusing (IEF). Traditionally, this method

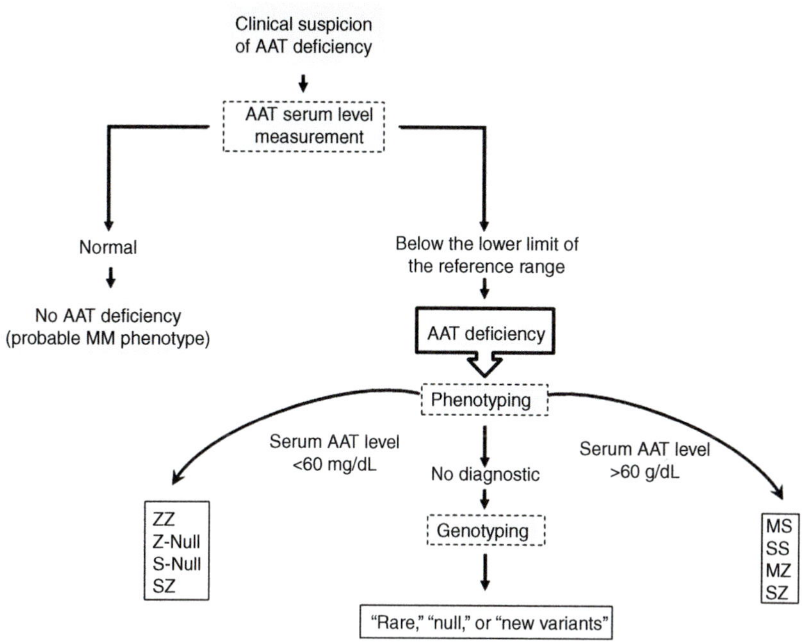

FIGURE 9.2 **Classical diagnostic testing algorithm of alpha-1 antitrypsin deficiency.** Traditionally, the diagnosis of deficit of alpha-1 antitrypsin is based on the quantification by nephelometry of the protein in serum or plasma, isoelectric focusing (IEF) in subjects with low levels (phenotyping), and sequencing of the gene (genotyping) if necessary as the last resort.

has been regarded as the gold standard in AAT diagnosis because IEF is a relatively inexpensive and safe test, despite the fact that the interpretation of results requires some expertise.

Predictably, if AAT values are above 35% of the lower reference limit, the most likely phenotypes may be MS, SS, MZ, or SZ. Only rarely will be necessary genotyping to correctly identify these alleles.

When the AAT serum levels are less than 35% of the lower reference limit, possible phenotypes may be Pi*Z (usually Pi*ZZ, but occasionally Pi*Z-Null), or exceptionally Pi*S-Null or Pi*SZ. Only rarely it may be necessary to perform gene sequencing of the SERPINA-1 gene to investigate null or rare variants (Fig. 9.2).

The use of DBS samples for AATD diagnosis has become widely available and has facilitated testing in recent years. AAT levels can be measured from DBS with moderate-to-good correlation with AAT serum levels [43]. If DBS is used instead of serum or plasma for IEF phenotyping, genotyping by allele-specific PCR methods allows identification of the most common AATD alleles: Pi*S and Pi*Z [40]. However, if AAT serum level and allele-specific PCR do not give consistent results or if confirmation of DBS samples findings is yet needed, the classical serum nephelometry and IEF are the methods of choice (Fig. 9.3).

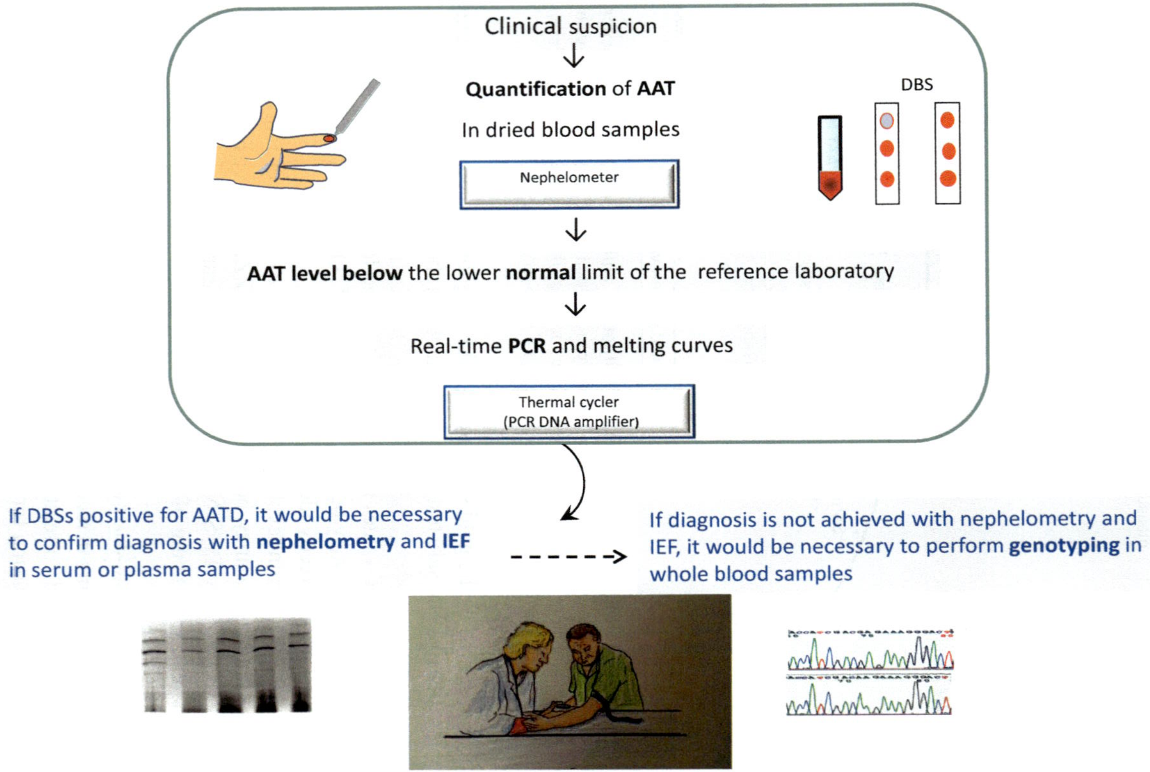

FIGURE 9.3 Alternative protocol for the diagnosis of alpha-1 antitrypsin deficiency using dried blood spot samples collected by fingerprick and dried on filter paper.

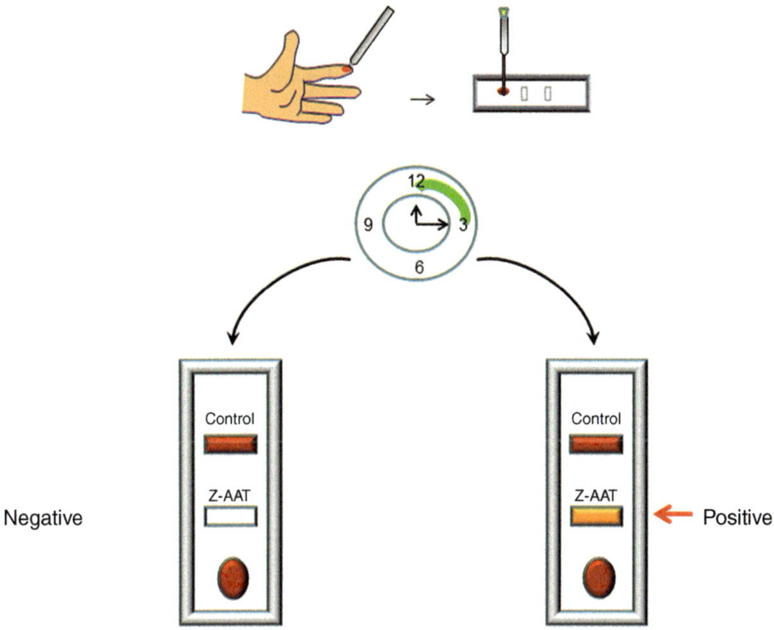

FIGURE 9.4 Rapid test to detect the Z-alpha-1 antitrypsin protein in fresh capillary blood samples collected by fingerprick. This small portable device can detect the Z protein in fresh blood of the patient. The result can be read at the top peephole in approximately 15–20 min, and reports about the presence or the Z protein, in both homozygotes ZZ and heterozygotes MZ, SZ, or Pi*Z-.

Gene sequencing of the SERPINA-1 gene is currently reserved for cases in which the serum level is low, and this finding cannot be explained by allele-specific PCR or IEF methods. The nucleotide obtained sequence should be compared with reference sequences of known mutations via Internet-based databases [40].

Recently, a new test for AATD based on the principle of a lateral flow assay [44] was introduced with a very manageable format, portable (approximate size 7 × 3 cm), compact, easy to use, and with a very fast interpretation of results (i.e., goes positive or negative in approximately 15–20 min). The test works with a Z protein-specific antibody, thus detecting only the Z protein (present in over 90% of cases with AATD) in a drop of blood obtained by puncture with a lancet in the tip of a finger (Fig. 9.4). As the test cannot differentiate between homozygous and heterozygous carriers of the Z mutation, further confirmatory analyses are need in positive cases [27].

REFERENCES

[1] American Thoracic Society/European Respiratory Society. American Thoracic Society/European Respiratory Society statement: standards for the diagnosis and management of individuals with alpha1-antitrypsin deficiency. Am J Respir Crit Care Med 2003;168:818–900.

[2] Stone H, Pye A, Stockley RA. Disease associations in alpha-1-antitrypsin deficiency. Respir Med 2014;108:338–43.

[3] de Serres F, Blanco I. Role of alpha-1 antitrypsin in human health and disease. J Intern Med 2014;276(4):311–35.

[4] Blanco I, Lipsker D, Lara B, Janciauskiene S. Neutrophilic panniculitis associated with alpha-1 antitrypsin deficiency: an update. Br J Dermatol 2016;174:753–62.

[5] DeMeo DL, Silverman EK. Alpha1-antitrypsin deficiency. 2: genetic aspects of alpha(1)-antitrypsin deficiency: phenotypes and genetic modifiers of emphysema risk. Thorax 2004;59:259–64.

[6] Green CE, Vayalapra S, Hampson JA, Mukherjee D, Stockley RA, Turner AM. PiSZ alpha-1 antitrypsin deficiency (AATD): pulmonary phenotype and prognosis relative to PiZZ AATD and PiMM COPD. Thorax 2015;70:939–45.

[7] Stoller JK, Lacbawan FL, Aboussouan LS. Alpha-1 antitrypsin deficiency. In: Pagon, RA, Adam, MP, Ardinger, HH, et al., editors. GeneReviews® [Internet]. Seattle, WA: University of Washington; 1993–2016. Available from: http://www.ncbi.nlm.nih.gov/books/NBK1519/

[8] Stoller JK, Aboussouan LS. Alpha1-antitrypsin deficiency. Lancet 2005;365:2225–36.

[9] Stoller JK, Aboussouan LS. A review of α1-antitrypsin deficiency. Am J Respir Crit Care Med 2012;185:246–59.

[10] Cox D, Levison H. Emphysema of early onset associated with a complete deficiency of alpha-1-antitrypsin (null homozygotes). Am Rev Respir Dis 1988;137:371–5.

[11] Fregonese L, Stolk J, Frants RR, Veldhuisen B. Alpha-1 antitrypsin Null mutations and severity of emphysema. Respir Med 2008;102:876–84.

[12] Turino GM, Barker AF, Brantly ML, Cohen AB, Connelly RP, Crystal RG, et al. Clinical features of individuals with PI*SZ phenotype of alpha-1 antitrypsin deficiency. Alpha-1 Antitrypsin Deficiency Registry Study Group. Am J Respir Crit Care Med 1996;154:1718–25.

[13] de Serres FJ. Alpha-1 antitrypsin deficiency is not a rare disease but a disease that is rarely diagnosed. Environ Health Perspect 2003;111:1851–4.

[14] Hutsebaut J, Janssens W, Louis R, Willersinn F, Stephenne X, Sokal E, et al. Activity of the alpha-1 antitrypsin deficiency registry in Belgium. COPD 2015;12(Suppl. 1):10–4.

[15] Bradi AC, Audisho N, Casey DK, Chapman KR. Alpha-1 antitrypsin deficiency in Canada: regional disparities in diagnosis and management. COPD 2015;12(Suppl. 1):15–21.

[16] Gauvain C, Mornex JF, Pison C, Cuvelier A, Balduyck M, Pujazon MC, et al. Health-related quality of life in patients with alpha-1 antitrypsin deficiency: the French experience. COPD 2015;12(Suppl. 1):46–51.

[17] Fähndrich S, Herr C, Greulich T, Seibert M, Lepper PM, Bernhard N, et al. Sex differences in alpha-1-antitrypsin deficiency lung disease-analysis from the German registry. COPD 2015;12(Suppl. 1):58–62.

[18] O'Brien ME, Pennycooke K, Carroll TP, Shum J, Fee LT, O'Connor C, et al. The impact of smoke exposure on the clinical phenotype of alpha-1 antitrypsin deficiency in Ireland: exploiting a national registry to understand a rare disease. COPD 2015;12(Suppl. 1):2–9.

[19] Luisetti M, Ferrarotti I, Corda L, Ottaviani S, Gatta N, Tinelli C, et al. Italian registry of patients with alpha-1 antitrypsin deficiency: general data and quality of life evaluation. COPD 2015;12(Suppl. 1):52–7.

[20] Stolk J, Fumagalli M, Viglio S, Iadarola P. Conductivity in exhaled breath condensate from subjects with emphysema and type ZZ alpha-1-antitrypsin deficiency. COPD 2015;12(Suppl. 1): 32–5.

[21] Chorostowska-Wynimko J, Struniawski R, Sliwinski P, Wajda B, Czajkowska-Malinowska M. The national alpha-1 antitrypsin deficiency registry in Poland. COPD 2015;12(Suppl. 1):22–6.

[22] Lara B, Miravitlles M. Spanish registry of patients with alpha-1 antitrypsin deficiency; comparison of the characteristics of PISZ and PIZZ individuals. COPD 2015;12(Suppl. 1):27–31.

[23] Piitulainen E, Tanash HA. Severe alpha 1-antitrypsin deficiency. COPD 2015;12(Suppl. 1):36–41.

[24] Stockley RA. Antitrypsin Deficiency Assessment and Programme for Treatment (ADAPT): the United Kingdom Registry. COPD 2015;12(Suppl. 1):63–8.

[25] de Serres FJ, Blanco I. Prevalence of α1-antitrypsin deficiency alleles PI*S and PI*Z worldwide and effective screening for each of the five phenotypic classes PI*MS, PI*MZ, PI*SS, PI*SZ, and PI*ZZ: a comprehensive review. Ther Adv Respir Dis 2012;6:277–95.

[26] Stoller JK, Sandhaus RA, Turino G, Dickson R, Rodgers K, Strange C. Delay in diagnosis of alpha1-antitrypsin deficiency: a continuing problem. Chest 2005;128:1989–94.

[27] Greulich T, Vogelmeier CF. Alpha-1-antitrypsin deficiency: increasing awareness and improving diagnosis. Ther Adv Respir Dis 2016;10:72–84.

[28] Greulich T, Ottaviani S, Bals R, Lepper PM, Vogelmeier C, Luisetti M, et al. Alpha1-antitrypsin deficiency—diagnostic testing and disease awareness in Germany and Italy. Respir Med 2013;107:1400–8.

[29] Köhnlein T, Janciauskiene S, Welte T. Diagnostic delay and clinical modifiers in alpha-1 antitrypsin deficiency. Ther Adv Respir Dis 2010;4:279–87.

[30] Campos MA, Wanner A, Zhang G, Sandhaus RA. Trends in the diagnosis of symptomatic patients with alpha1-antitrypsin deficiency between 1968 and 2003. Chest 2005;128:1179–86.

[31] Rahaghi F, Ortega I, Rahaghi N, Oliveira E, Ramirez J, Smolley L, et al. Physician alert suggesting alpha-1 antitrypsin deficiency testing in pulmonary function test (PFT) results. COPD 2009;6:26–30.

[32] Molina J, Flor X, García R, Timiraos R, Tirado Conde G, Miravitlles M. The IDDEA project: a strategy for the detection of alpha-1 antitrypsin deficiency in COPD patients in the primary care setting. Ther Adv Respir Dis 2011;5:237–43.

[33] Alpha 1-antitrypsin deficiency: memorandum from a WHO meeting. Bull World Health Organ 1997;75:397–415.

[34] Stoller J, Fromer L, Brantly M, Stocks J, Strange C. Primary care diagnosis of alpha-1 antitrypsin deficiency: issues and opportunities. Cleve Clin J Med 2007;74:869–74.

[35] Luisetti M, Seersholm N. Alpha1-antitrypsin deficiency. 1: epidemiology of alpha1-antitrypsin deficiency. Thorax 2004;59:164–9.

[36] Yurkiewicz IR, Korf BR, Lehmann LS. Prenatal whole-genome sequencing—is the quest to know a fetus's future ethical? N Engl J Med 2014;370:195–7.

[37] Cox DW. Prenatal diagnosis for alpha1-antitrypsin deficiency. Prenat Diagn 2004;24:468–70.

[38] Herzenberg LA, Bianchi DW, Schröder J, Cann HM, Iverson GM. Fetal cells in the blood of pregnant women: detection and enrichment by fluorescence-activated cell sorting. Proc Natl Acad Sci USA 1979;76:1453–5.

[39] Alfirevic Z, Mujezinovic F, Sundberg K. Amniocentesis and chorionic villus sampling for prenatal diagnosis (Review). Cochrane Database Syst Rev 2003;(3). Doi: 10.1002/14651858.CD003252.

[40] Miravitlles M, Herr C, Ferrarotti I, Jardi R, Rodriguez-Frias F, Luisetti M, et al. Laboratory testing of individuals with severe alpha1-antitrypsin deficiency in three European centres. Eur Respir J 2010;35:960–8.

[41] Bals R, Koczulla R, Kotke V, Andress J, Blackert K, Vogelmeier C. Identification of individuals with alpha-1-antitrypsin deficiency by a targeted screening program. Respir Med 2007;101:1708–14.

[42] Bornhorst JA, Greene DN, Ashwood ER, Grenache DG. α1-Antitrypsin phenotypes and associated serum protein concentrations in a large clinical population. Chest 2013;143:1000–8.
[43] Costa X, Jardi R, Rodriguez F, Miravitlles M, Cotrina M, Gonzalez C, et al. Simple method for alpha1-antitrypsin deficiency screening by use of dried blood spot specimens. Eur Respir J 2000;15:1111–5.
[44] Yetisen AK, Akram MS, Lowe CR. Paper-based microfluidic point-of-care diagnostic devices. Lab Chip 2013;13:2210–51.

Chapter 10

Liver Disease Associated With Alpha-1 Antitrypsin Deficiency

10.1 INTRODUCTION

10.1.1 Discovery of the Hepatic Disease Associated With Alpha-1 Antitrypsin Deficiency

The alpha-1 antitrypsin (AAT) deficiency (AATD) history begins in Malmoe in the early years of the decade of the 1960s, when Laurell and his collaborator Eriksson reported the first series in the world of patients with AATD and lung emphysema [1–3].

In 1962, just a year prior to his first communication of AATD-related lung emphysema, Eriksson himself also found an AATD patient with advanced liver cirrhosis, but he interpreted that this patient had developed an AATD secondary to his severe liver dysfunction [4].

Few years later (1969), the association between liver disease and AATD was first documented in children liver biopsy samples by Sharp and colleagues, in Minnesota [5]. By electron microscopy, this author found an amorphous proteinaceous material within the lumen of the endoplasmic reticulum (ER) of hepatocytes, and using the light microscope, he found that this amorphous material presented as eosinophilic globules, most prominently in periportal hepatocytes, easily seen with PAS (Periodic Acid Schiff) staining [6].

The pioneering discoveries of Sharp were later explained in the 1990s, when Lomas in Cambridge showed that the mutant Z AAT forms intrahepatic polymers due to the insertion of the reactive center loop of one molecule into the opened B-sheet of an adjoining molecule, discovering that polymerization is a critical step in the pathophysiology of the disease [7].

10.1.2 Pathophysiological Mechanisms

As explained in Chapter 5, the amino acids of the mutated proteins are improperly assembled in the rough ER where they aggregate forming linear polymers. The accumulation of mutated proteins in the ER induces mitochondrial damage, apoptotic signals from mitochondria, and activation of autophagic pathways. Continued cellular stress and subsequent repair lead to liver fibrosis and cirrhosis of the injured liver. Other factors, such as systemic inflammation, exogenous toxic substances, and infections, can act as a second factor that increases the synthesis of mutated proteins and the amount of polymerized proteins [8].

In general, Pi*Z homozygotes, and homozygotes or compound heterozygotes with rare deficiency alleles, such as Mmalton, Mduarte, Siiyama, and so on, are at high risk of developing liver disease [9]. This risk is greatest in the neonatal age and after the 50 years, but it is currently unknown why only a relatively small proportion of Pi*ZZ individuals progress to cirrhosis.

10.1.3 Clinical Presentation

The clinical presentation of liver disease in people with AATD is highly variable, ranging from self-limiting subclinical hepatitis to chronic hepatitis evolving to cirrhosis (clinically well tolerated or rapidly evolving to fulminant hepatic failure), and hepatocellular carcinoma, generally developed in elderly people with either cirrhotic or noncirrhotic liver [10].

10.2 CHILDHOOD-ONSET LIVER DISEASE

10.2.1 Clinical Symptoms and Signs of Presentation

In infants and children, AATD is the most common genetic cause of liver disease [8]. The typical clinical presentation in the neonatal period is an intrahepatic cholestasis (also known as neonatal hepatitis syndrome) with prolonged obstructive jaundice, acholic stools, hyperchromic urine, abdominal bloating, itching, anorexia, poor weight gain, hepatomegaly and splenomegaly (Fig. 10.1), serum elevated levels of total and conjugated bilirubin, alanine transaminase (ALT), aspartate transaminase (AST), gamma-glutamyltransferase (GGT), hypoalbuminemia, and vitamin K deficiency [9].

10.2.2 Pathology

The pathological changes in liver biopsy are highly variable, including steatosis, fibrosis, hepatocellular necrosis, giant cell transformation, lobular inflammation, intrahepatic cholestasis, bile duct poverty, or bile duct proliferation in portal areas. Occasionally, in infants globular eosinophilic inclusions may be detected in some hepatocytes, as well as in bile duct epithelium, although they are not usually observed before the age of 12 weeks, and even above this age these inclusions may be fine, granular, and difficult to identify [10].

In more advanced stages of childhood, PAS+ and diastase-resistant globules may be detected, but remarkably these globules are not absolutely specific of AATD and should be interpreted with caution, as they can be seen in other liver diseases [11]. In fact, some PAS+ and diastase-resistant globules (called type 1) are periportal in location, stain positively with the specific AAT immunoperoxidase method, and are associated with AATD. However, other globules,

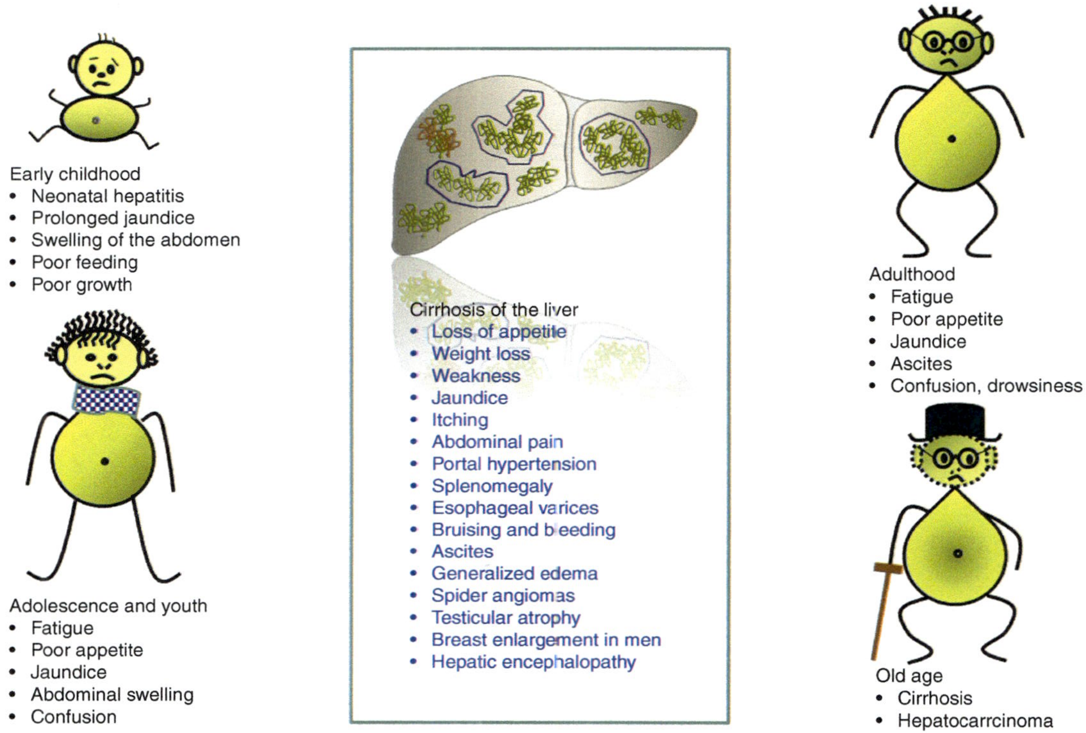

FIGURE 10.1 Alpha-1 antitrypsin deficiency liver disease: clinical symptoms and signs. Clinically apparent liver disease associated with alpha-1 antitrypsin deficiency (typically liver cirrhosis) has a higher incicence in childhood and old age, but may occur at any age. Some adults and elderly persons can also develop liver carcinomas within cirrhotic or noncirrhotic livers The symptoms do not differ from those generated by other pathologies causing chronic hepatitis or cirrhosis.

called type 2, are located in the centrilobular region of the liver, or at the edge of the central ischemic zone, and they do not stain with the immunoperoxidase technique. Type 2 globules are nonspecific and can be seen in any liver disease accompanied by sinusoidal congestion, hemorrhage, necrosis, portal fibrosis, or anoxia. Therefore, immunocytochemical methods are useful in separating the two types of PAS-positive diastase-resistant globules [12]. In latest years, several monoclonal antibodies that specifically identify AAT polymers have been developed [13,14]. Interestingly, the number and size of liver globules increase with age [11].

10.2.3 Natural History

The natural history of the AATD–liver disease in previous stages of life to adulthood is well known, thanks to a valuable study led by Sveger (Department of Pediatrics, University Hospital of Malmoe, Sweden), who screened for AATD 200,000 newborns (more than 95% of all borns between 1972 and 1974 in hospitals across Sweden). A serum AAT cutoff of 40% of the reference value was used in this study, detecting all ZZ, Z-null, and 1/3 of SZ types but not the MZ, two-thirds of SZ, and the rest of phenotypes expressing AAT serum levels over 40%. With this methodology, 190 infants with severe AATD were identified, that is, 120 Pi*ZZ, 48 Pi*SZ, 2 Pi*Z-null, 1 Pi*S-null, and 19 not identified phenotypes. The subsequent detection of new cases increased the number of Pi*ZZ and Pi*SZ to 127 and 54, respectively [15].

The evolution of this Swedish cohort in the early years of life was reported in 1976 and 1978 by Sveger [15,16], and their clinical and analytical situation at 16 and 18 by Sveger and Eriksson in 1995 [17]. More recently in 2015, Tanash et al. reported the health status, and the lung and liver function of these patients at the age of 34 years [18]. The data of these studies are presented together in Table 10.1.

In short, in the first year of the total 127 Pi*ZZ infants, 14 (11%) presented neonatal cholestasis with jaundice, and 8 (6%) anicteric hepatitis (i.e., hepatitis without jaundice), with hepatosplenomegaly, growth retardation, and abnormal liver enzymes at the time of the medical examination (within 3–6 months). In general, the appearance of clinical symptoms ranged from 1 week to 12 months, although in most patients the clinical and laboratory abnormalities were evident in the first 4 months. In the second year of life, 3 of the 14 cases with neonatal cholestasis progressed to cirrhosis, 2 developed hepatomegaly and 1 developed a fatty liver. Between 2 and 4 years, one of the three children with cirrhosis died by aplastic anaemia; the other two children followed with their cirrhosis; and hepatomegaly persisted in one of the two cases with an enlarged liver. Between 4 and 8 years, two of the first children with cholestasis evolved into cirrhosis died from liver cirrhosis. From age 8, none of the 11 survivors with cholestasis neither any of the 8 with anicteric hepatitis showed clinical signs of liver disease, although most continued with high serum levels of ALT and GGT. At the

TABLE 10.1 Natural History From Birth to Age 34 of 127 Pi*Z and 54 Pi*SZ Alpha-1 Antitrypsin Deficiency Swedish Subjects

	127 Pi*ZZ		54 Pi*SZ
Chronology	**105 Subclinical hepatitis (83%)**	**22 Neonatal hepatitis (17%)**	**7 Subclinical hepatitis**
3–12 months	Slight elevation without clinical expression of ALT or GGT in >80%	• Neonatal cholestatic hepatitis with jaundice in 14 cases (11%) • Anicteric hepatitis with laboratory abnormalities and hepatosplenomegaly in 8 cases (6%)	Slight elevation without clinical expression of ALT and GGT in 13%
2 years	Persists slight elevation without clinical expression of ALT in 51 cases (49%)	• 3 Hepatitis progressed to liver cirrhosis • 2 Hepatitis developed hepatomegaly • 1 Hepatitis developed fatty liver (liver biopsy+)	Subclinical elevation of ALT in 23%
3 years		• 1 Child with cirrhosis died from aplastic anemia • Cirrhosis persisted in 2 cases • Hepatomegaly persisted in 1 case	
4 years	Slight elevation of ALT in 48 cases	• The 2 remaining cirrhotic patients died from liver failure	Subclinical elevation of ALT in 2%
8 years	Slight elevation of ALT in 44 cases and GGT in 5 cases	• Laboratory abnormalities without clinical liver disease in the 19 survivors	Subclinical elevation of ALT in 2%
12 years	Slight elevation of ALT (18 cases) and GGT (4 cases)	• Laboratory abnormalities (i.e., elevation of ALT and GGT) without clinical expression in 3 survivors	Subclinical elevation of ALT in 2%
16 years	Slight elevation of ALT (19 cases) and GGT (9 cases)	• Slight elevation of ALT and GGT without clinical expression in 2 survivors	Subclinical elevation of ALT and GGT in 4%
18 years	Slight elevation of ALT (13 cases) and GGT (9 cases)	• Slight elevation of ALT and GGT without clinical expression in 2 survivors	Subclinical elevation of ALT and GGT in 9%
34 years	Liver enzymes within normal range	• Liver enzymes within normal range	Liver enzymes within normal range

ALT, Alanine aminotransferase; GGT, gamma glutamyltransferase.

age of 12 years, 20% of survivors with neonatal cholestasis and 14% of those with anicteric hepatitis showed elevated transaminases without clinically symptoms of liver disease [15–17].

In 2015, the Malmo team was able to assess 116 Pi*ZZ and 48 Pi*SZ of the initial cohort after 34 years, and they did not find significant alterations in liver enzymes. In addition, most of the 105 Pi*ZZ infants belonging to the group of newborns without clinical manifestations showed subclinical neonatal hepatitis, documented by asymptomatic mild and transient elevation of ALT and/or GGT, which gradually normalized [18].

In summary, the majority of the Pi*ZZ Swedish children who developed neonatal cholestasis had a favorable outcome, with a risk of death from childhood cirrhosis of 2%–3%. In addition, most children who presented altered liver function test without clinical manifestations during the first year of life normalized after 12 years.

On the other hand, liver involvement in the group of 54 Pi*SZ newborns was benign and less frequent than in the ZZ group. At age 34, all of these Pi*SZ were asymptomatic and liver enzymes normalized [15–18].

This observation coincides with what was observed by Pittschieler in Bozen (Italy) in 60 children with SZ and MZ phenotypes who had subclinical elevation of liver enzymes at birth, and between 5 and 10 years spontaneously returned to normal values [19].

Apart from the prospective Swedish study, the results of other observational series have been reported [20–25]. Many of these are retrospective studies, and their results are more pessimistic than those provided by the Swedish study (Table 10.2).

It is not clear why the liver inflammation of children with neonatal hepatitis associated with AATD can lead to cirrhosis, or to stabilize and even disappear in childhood–adolescence. Nevertheless, several poor prognosis factors have been identified, including elevation of serum transaminases 3 times above normal; persistent hypertransaminasemia and prolonged coagulation time; delayed resolution of jaundice; persistent splenomegaly; persistent high levels of bile acids in urine; liver biopsy with neo ductal proliferation and portal fibrosis [9,25].

Although some initial studies reported a high prevalence of the hepatitis B antigen in the blood of newborns with neonatal hepatitis syndrome and also in their parents, suggesting this could be an important trigger factor that would lead to cirrhosis [27,28], subsequent studies have not confirmed this association.

The fate of the children with Pi*ZZ phenotype follows that of the preceding siblings with the same phenotype in a high percentage of cases. For example, the risk for a Pi*ZZ fetus to develop severe liver disease has been estimated at 13% if a previous Pi*ZZ sib had no liver disease or it resolved during early childhood, and a risk of 40% if a previous Pi*ZZ sib had developed severe liver disease [29].

TABLE 10.2 Evolution of ZZ Alpha-1 Antitrypsin Deficiency-Related Neonatal Hepatitis in Five Selected Series

Author, country, and year [reference]	Sample compositior	Clinical presentation	Clinical evolution
Sveger, Sweden, 1976, 2015 [15–18]	127 Infants followed up to 34 years (prospective in general population)	- 14 Neonatal cholestatic hepatitis - 8 Neonatal anicteric hepatitis with hepatosplenomegaly	- 3 Cholestatic hepatitis evolved to cirrhosis and died before age 7 years At age 34, liver enzymes normalized in all survivors
Moroz, Canada, 1976 [20]	18 Newborns	- 14 Cholestatic hepatitis - 4 Anicteric hepatitis with hepatomegaly	- 7 Progressed to cirrhosis and 4 of these died - In 11 with more than 13 years liver disease persisted with mild laboratory and histological abnormalities
Psacharopoulos, UK, 1983 [21]	136 Children	- 74 Chronic liver disease	- 20 Deaths from cirrhosis before age 17 - 39 Continued with hepatitis or cirrhosis - 15 Has both clinical and biochemical remission
Nebbia, France, 1983 [22] and Alagille, France, 1984 [23]	45 Children	- 45 Neonatal cholestasis	- 25 Developed liver cirrhosis (8 died between 1 and 14 years of age) - 20 Followed a favorable course, but some asymptomatic subjects presented portal fibrosis on liver biopsy
Ghishan, USA, 1988 [24]	18 Children	- 15 Neonatal cholestasis - 3 Anicteric hepatitis	- 3 Were transplanted due to cirrhosis at 3, 4, and 7 years - 2 Died at 4 months and 3 years for cirrhosis - 2 Anicteric hepatitis evolved favorably
Hussain, UK, 1991 [25]	192 New-borns	- 178 Neonatal cholestasis - 14 Cirrhosis	- 27 Deaths with an average age of 6 years (range: 1–17) due to complications of cirrhosis
Volpert, USA, 2000 [26]	44 Children	- 17 Cirrhosis	- 2 Underwent liver transplantation - The rest are leading relatively healthy lives for 4–23 years after diagnosis

10.3 ADOLESCENCE AND YOUTH-ONSET LIVER DISEASE

In adolescents and young people, AATD-related liver disease may present as an asymptomatic subclinical hepatitis casually detected by a slight elevation of transaminases. However, there are also some patients with severe liver disease in the first months or years of life, who may enter a period called "honeymoon" with few symptoms or signs and normal growth, before entering a period of disease progression in adolescence. Other with a delay in growth associated with anorexia or hepatomegaly with or without splenomegaly, may be detected during a routine medical visit. Occasionally, some subjects may present with ascites, gastrointestinal bleeding, or liver failure (Fig. 10.1). The incidence of these problems is not well established, but often most of them may have not been previously detected [30].

Histologically, the characteristic feature of the AATD–liver injury is an accumulation of Z protein polymers into the cytoplasm of hepatocytes, as positive to PAS staining and diastase-resistant globules (Fig. 10.2).

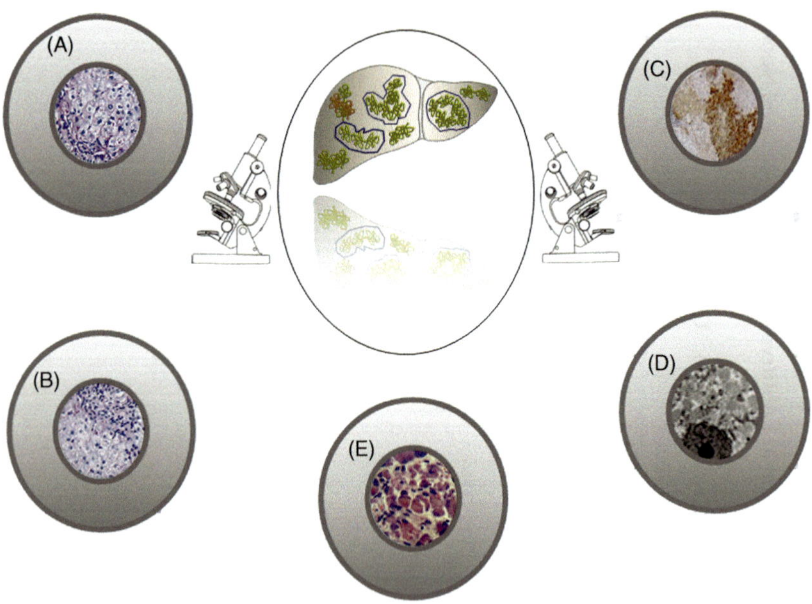

FIGURE 10.2 Alpha-1 antitrypsin deficiency liver disease: pathological changes. Histologically, the most characteristic feature of the alpha-1 antitrypsin deficiency liver damage is the presence of polymers of the mutated protein into the cytoplasm of hepatocytes, as PAS (Periodic Acid Schiff)-positive and diastase-resistant cytoplasmic eosinophilic (*pink*) globules in periportal hepatocytes (A, B). Electron microscopy shows granular material within a dilated rough endoplasmic reticulum (D). As PAS staining is not absolutely specific, immunohistochemical staining (*intense brown* staining) can provide a more accurate diagnosis (C). Dysplastic hepatocytes with pleomorphic and prominent nuclei are present in alpha-1 antitrypsin deficiency-related hepatocellular carcinomas (E).

10.4 ADULT-ONSET LIVER DISEASE

10.4.1 Clinical Symptoms and Signs

AATD is one of the most common causes of liver cirrhosis in adults, behind alcohol abuse, viral hepatitis, and chronic cholangitis. Over one-third of genetically susceptible adult patients with Pi*ZZ genotype develops liver cirrhosis, and autopsy studies suggest that the prevalence of liver disease may be as high as 40% in older individuals who have never smoked reaching advanced ages without developing COPD [11].

Liver disease in adults may occur in the absence of a history of neonatal or childhood liver disease, and it is more frequent in men than women. In a pioneering study of 246 Swedish Pi*ZZ with cirrhosis, it was found that 104 of them (2%) were aged between 20 and 50 years, and the rest (142, 19%) exceeded 50 years [31]. A subsequent study in Canada also demonstrated that the risk for the development of liver disease was relatively high for men between 51 and 60 years, and again the risk for women was lower than that for men [32].

Liver involvement may be manifested by an asymptomatic hepatitis with elevated transaminases casually detected in a medical consultation. Other cases present with advanced symptoms and signs of decompensated cirrhosis, no different from those caused by alcohol consumption or viral hepatitis, consisting of a mixture of clinical manifestations, such as loss of appetite, weight loss, weakness, jaundice, itching, portal hypertension, hepatosplenomegaly, abdominal pain, esophageal varices, ascites, anasarca, slow mental functioning, and so on (Fig. 10.1). The progression of cirrhosis in AATD is usually slow, but some patients develop a rapid disease with need of liver transplantation in the short term [11].

Histologically, the characteristic features of liver cirrhosis are regenerative nodules surrounded by fibrous bands leading to portal hypertension, along with a remarkable accumulation of PAS-positive staining and diastase-resistant globules, visualized as bright pink globules of various sizes (Fig. 10.2).

10.4.2 Liver Carcinoma

There is an increased incidence of hepatocellular carcinomas, and also, although less frequently, of cholangiocarcinomas and hepatocolangiocarcinomas in Pi*ZZ subjects with well-established cirrhosis and occasionally with noncirrhotic liver. The estimated risk for carcinomas is estimated at 2%–3%, although this percentage rises to 29% in series of autopsies [33–37]. The risk of developing cancer is especially significant in males, with no evidence that alcoholism and viral hepatitis are additional risk factors. The high incidence of liver carcinomas has been attributed to the polymer-overloaded hepatocytes which send strong and continuous signals to the hepatocytes without load of Z proteins, stimulating their cell cycle speed [11].

10.5 LIVER DISEASE AND HETEROZYGOSITY

The association between AATD heterozygosity (i.e., subjects having a single Pi*Z allele) and risk of developing a chronic liver disease is controversial by the lack of studies with sufficient statistical power to get reliable conclusions.

With the available data it has been estimated that the odds ratio of developing cirrhosis by MZ heterozygotes varies between 1.8 and 3.1, and is higher in men [38–41]. However, the risk is much lower (3%) when compared with that of ZZ homozygotes (30%). In a series of 641 adults presenting with advanced chronic liver failure before orthotopic liver transplantation, a greater number of MZ heterozygotes (8.4%) was observed than that were reported in the general population (2%–4%) [41]. Chronic liver disease in AATD heterozygotes is associated with a high prevalence of viral infection, and this infection, rather than AATD alone, may be the cause of the liver disease in such patients [39]. Therefore, contrary to what happens in the ZZ homozygotes, it seems that chronic alcohol consumption and hepatitis B and C are necessary additional factors for the development of liver cirrhosis in MZ heterozygotes [40].

The S variant has an increased susceptibility to polymerization, although this is marginal compared with the more conformationally unstable Z variant. The SZ genotypes produce S and Z proteins that form heteropolymers into hepatocytes able to induce the development of liver cirrhosis [42]. However, the available knowledge about this aspect is scarce and limited to isolated cases and small series. On the contrary, there is no evidence that the phenotypes MS and SS promote the development of liver disease [11].

Although the risk for liver disease development associated with some rare alleles (e.g., Pi*Mmalton, Pi*Mduarte) is unknown, its marked propensity to polymerize and form PAS-D-positive inclusion bodies into hepatocytes makes it very likely that its risk is similar to that of Pi*ZZ. For its rare prevalence, there are not enough clinical data to assess with certainty the risk of these alleles; nevertheless, case reports describe the association of some rare phenotypes with hepatic fibrosis, cirrhosis, dysplasia, and hepatocellular cancer [11,43].

10.6 TREATMENT

There is no specific treatment for the liver disease associated with AATD. Augmentation therapy is not useful (but not harmful) because the liver damage is related to the accumulation of mutant proteins within hepatocytes, and not for lack of circulating antiprotease activity [11].

The goals of treatment are to slow the progression of scar tissue and to prevent or treat symptoms and complications of cirrhosis, such as bleeding, ascites, pruritus, malnutrition, vitamin deficiencies, growth disorders, infections, malignancies, and so on. In general, the therapeutic measures available are the same as those applied in other liver diseases of different etiologies leading to liver cirrhosis. Close monitoring of the progression of the disease is necessary to transfer patients to centers with liver orthotopic transplantation services when indicated.

For those patients who develop decompensated cirrhosis or hepatocellular carcinoma, liver transplantation replaces the damaged liver and corrects the serum deficiency of AAT. It must be taken into account that the recipient of the new organ changes its former ZZ phenotype by that of the donor (usually MM), although the genotype does not change and continues being a Pi*ZZ.

In pediatric liver transplant centers, AATD is the second most common cause of transplantation, only surpassed by biliary atresia. Overall, children who survive more than 1 year after transplantation have a reasonably good prognosis, although long-term complications of immunosuppression can be expected [44].

REFERENCES

[1] Laurell CB, Eriksson S. The electrophoretic alpha1-globulin pattern of serum in alpha1-antitrypsin deficiency. Scand J Clin Lab Invest 1963;15:132–40.

[2] Eriksson S. Pulmonary emphysema and alpha-l-antitrypsin deficiency. Acta Med Scand 1964;175:197–205.

[3] Laurell CB, Eriksson S. The serum alpha-l-antitrypsin in families with hypo-alpha-l-antitrypsinemia. Clin Chim Acta 1965;11:395–8.

[4] Janciauskiene SM, Bals R, Koczulla R, Vogelmeier C, Köhnlein T, Welte T. The discovery of α1-antitrypsin and its role in health and disease. Respir Med 2011;105:1129–39.

[5] Sharp HL, Bridges RA, Krivit W, Freier EF. Cirrhosis associated with alpha-1-antitrypsin deficiency: a previously unrecognized inherited disorder. J Lab Clin Med 1969;73:934–9.

[6] Sharp HL. History of the first description of childhood liver disease in AATD. COPD 2013;10(Suppl. 1):13–6.

[7] Lomas DA. Loop-sheet polymerization: the structural basis of Z alpha 1-antitrypsin accumulation in the liver. Clin Sci (Lond) 1994;86:489–95.

[8] Teckman JH. α1-Antitrypsin deficiency in childhood. Semin Liver Dis 2007;27:274–81.

[9] Bals R. Alpha-1-antitrypsin deficiency. Best Pract Res Clin Gastroeneterol 2010;24:629–33.

[10] Cutz E, Cox DW. Alpha-1 antitrypsin deficiency: the spectrum of pathology and pathophysiology. Perspect Pediatr Pathol 1979;5:1–39.

[11] American Thoracic Society/European Respiratory Society. American Thoracic Society/European Respiratory Society statement: standards for the diagnosis and management of individuals with alpha1-antitrypsin deficiency. Am J Respir Crit Care Med 2003;168:818–900.

[12] Qizilbash A, Young-Pong O. Alpha 1 antitrypsin liver disease differential diagnosis of PAS-positive, diastase-resistant globules in liver cells. Am J Clin Pathol 1983;79:697–702.

[13] Janciauskiene S, Eriksson S, Callea F, Mallya M, Zhou A, Seyama K, et al. Differential detection of PAS-positive inclusions formed by the Z, Siiyama, and Mmalton variants of alpha1-antitrypsin. Hepatology 2004;40:1203–10.

[14] Irving JA, Miranda E, Haq I, Perez J, Kotov VR, Faull SV, et al. An antibody raised against a pathogenic serpin variant induces mutant-like behaviour in the wild-type protein. Biochem J 2015;468:99–108.

[15] Sveger T. Liver disease in alpha 1-antitrypsin deficiency detected by screening of 200,000 infants. N Engl J Med 1976;294:1316–21.

[16] Sveger T. The natural history of liver disease in alpha 1 antitrypsin deficient children. Acta Paediatr Scand 1988;77:847–51.
[17] Sveger T, Eriksson S. The liver in adolescents with alpha 1-antitrypsin deficiency. Hepatology 1995;22:274–81.
[18] Tanash HA, Nystedt-Düzakin M, Cano Montero L, Sveger T, Piitulainen E. The Swedish alpha 1-antitrypsin screening study: health status, lung and liver function at age 34. Ann Am Thorac Soc 2015;12:807–12.
[19] Pittschieler K. Liver involvement in alpha1-antitrypsin-deficient phenotypes PiSZ and PiMZ. Acta Paediatr 2002;91:239–40.
[20] Moroz SP, Cutz E, Cox DW, Sass-Kortsak A. Liver disease associated with alpha1-antitrypsin deficiency in childhood. J Pediatr 1976;88:19–25.
[21] Psacharopoulos HT, Mowat AP, Cook PJ, Carlile PA, Portmann B, Rodeck CH. Outcome of liver disease associated with alpha 1 antitrypsin deficiency (PiZ). Implications for genetic counselling and antenatal diagnosis. Arch Dis Child 1983;58:882–7.
[22] Nebbia G, Hadchouel M, Odievre M, Alagille D. Early assessment of evolution of liver disease associated with alpha1-antitrypsin deficiency in childhood. J Pediatr 1983;102:661–5.
[23] Alagille D. Alpha-1-antitrypsin deficiency. Hepatology 1984;4:11S–4S.
[24] Ghishan FK, Greene HL. Liver disease in children with PiZZ alpha1-antitrypsin deficiency. Hepatology 1988;8:307–10.
[25] Hussain M, Mieli-Vergani G, Mowat AP. Alpha 1-antitrypsin deficiency and liver disease: clinical presentation, diagnosis and treatment. J Inherit Metab Dis 1991;14:497–511.
[26] Volpert D, Molleston JP, Perlmutter DH. Alpha-1 antitrypsin deficiency-associated liver disease progresses slowly in some children. J Pediatr Gastroenterol Nutr 2000;31:258–63.
[27] Porter Ca, Mowat AP, Cook PJL, Haynes DWG, Shilkin KB, Williams R. α1-Antitrypsin deficiency and neonatal hepatitis. Br Med J 1972;3:435–9.
[28] Cottrall K, Cook PJ, Mowat AP. Neonatal hepatitis syndrome and alpha1-antitrypsin deficiency: an epidemiological study in south-east England. Postgrad Med J 1974;50:376–80.
[29] Cox DW, Mansfield T. Prenatal diagnosis of alpha 1 antitrypsin deficiency and estimates of fetal risk for disease. J Med Genet 1987;24:52–9.
[30] Fairbanks KD, Tavill AS. Liver disease in alpha 1-antitrypsin deficiency: a review. Am J Gastroenterol 2008;103:2136–41.
[31] Larsson C. Natural history and life expectancy in severe alpha1-antitrypsin deficiency PiZ. Acta Med Scand 1978;294:345–51.
[32] Cox DW, Smyth S. Risk for liver disease in adults with alpha1-antitrypsin. Am J Med 1983;74:221–7.
[33] Eriksson S, Carlson J, Velez R. Risk of cirrhosis and primary liver cancer in alpha 1-antitrypsin deficiency. N Engl J Med 1986;314:736–9.
[34] Eriksson S. Alpha1-antitrypsin deficiency and liver cirrhosis in adults: an analysis of 35 Swedish autopsied cases. Acta Med Scand 1987;221:461–7.
[35] Elzouki AN, Eriksson S. Risk of hepatobiliary disease in adults with severe alpha 1-antitrypsin deficiency (PiZZ): an additional risk factor for cirrhosis and hepatocellular carcinoma? Eur J Gastroenterol Hepatol 1996;8:989–94.
[36] Palmer PE, Wolfe HJ. Alpha1-antitrypsin deposition in primary hepatic carcinoma. Arch Pathol Lab Med 1976;100:232–5.
[37] Reintoft I, Hägerstrand IE. Does the Z gene variant of alpha1-antitrypsin predispose to hepatic carcinoma? Hum Pathol 1979;10:419–24.
[38] Bell H, Schrumpf E, Fagerhol MK. Heterozygous MZ alpha1-antitrypsin deficiency in adults with chronic liver disease. Scand J Gastroenterol 1990;25:788–92.

[39] Propst T, Propst A, Dietze O, Judmaier G, Braunsteiner H, Vogel W. High prevalence of viral infection in adults with homozygous and heterozygous alpha1-antitrypsin deficiency and chronic liver disease. Ann Intern Med 1992;117:641–5.

[40] Eigenbrodt ML, McCashland TM, Dy RM, Clark J, Galati J. Heterozygous alpha1-antitrypsin phenotypes in patients with end stage liver disease. Am J Gastroenterol 1997;92:602–7.

[41] Graziadei IW, Joseph JJ, Wiesner RH, Therneau TM, Batts KP, Porayko MK. Increased risk of chronic liver failure in adults with heterozygous alpha1-antitrypsin deficiency. Hepatology 1998;28:1058–63.

[42] Mahadeva R, Chang WS, Dafforn TR, Oakley DJ, Foreman RC, Calvin J, et al. Heteropolymerization of S, I, and Z α1-antitrypsin and liver cirrhosis. J Clin Invest 1999;103:999–1006.

[43] Reid CL, Wiener GJ, Cox DW, Richter JE, Geisinger KR. Diffuse hepatocellular dysplasia and carcinoma associated with the Mmalton variant of alpha 1-antitrypsin. Gastroenterology 1987;93:181–7.

[44] Migliazza L, López Santamaría M, Murcia J, Gamez M, Clavijo J, Camarena C, et al. Long-term survival expectancy after liver transplantation in children. J Pediatr Surg 2000;35:5–7.

Chapter 11

Respiratory Manifestations of the Alpha-1 Antitrypsin Deficiency

11.1 INTRODUCTION

Many persons with severe alpha-1 antitrypsin (AAT) deficiency (AATD) do not have significant clinical manifestations of disease or lung function impairment throughout life [1], and among those who develop chronic obstructive pulmonary disease (COPD) there are a marked variability in the severity and age of onset of respiratory symptoms [2]. This variability depends on the lifetime tobacco smoking consumption and exposure to other environmental pollutants, the degree of bronchial hyperresponsiveness, the frequency and severity of lower respiratory tract infections, and some not well-known familial factors that contribute to a more severe clinical course [3].

During the first years of life, the only important threat for the health of AATD individuals is the possibility of developing liver diseases. Respiratory symptoms did not appear until adulthood, as shown by a study conducted in 150 adolescents (103 Pi*Z, 1 Pi*Z-, 1 Pi*S-, and 45 Pi*SZ) diagnosed in a Swedish nationwide neonatal screening and followed prospectively. At 16 years of age, these adolescents were asked to participate in an extensive lung function study. The conclusion was that children with AATD have a favorable prognosis and normal lung development up to 16 years of age, and that antismoking advice was reasonably successful, as only 3% of those answering the questionnaire had started to smoke [4]. More recently, another survey on 116 Pi*ZZ and 48 Pi*SZ from the aforementioned Swedish cohort corroborated the previous results finding no differences in lung function or symptoms between subjects with AATD and control subjects, but smoking frequency was significantly lower among the subjects with AATD at age 34 years [5].

11.2 CHILDHOOD-ONSET LUNG DISEASE

11.2.1 COPD

Although there have been reported isolated cases of COPD in children with AATD [6,7], this is an extremely rare occurrence which may result from the coexistence of other unidentified genetic or environmental factors affecting lungs [8].

Sometimes AATD is casually discovered by the absence of alpha-1 band in a protein electrophoresis, through neonatal screening, or in studies of relatives of index cases with AATD, but generally, if AATD has not debuted early with a neonatal hepatitis, this condition can go completely unnoticed in most cases for many years or even for a lifetime.

An example of the limited clinical repercussion of this genetic condition in childhood is reflected by the clinical course of all the 22 adolescents with homozygous Pi*Z genotype, whose condition was detected through neonatal screening in the early 1970s in Portland (USA), and had normal lung volumes, expiratory flow rates, and diffusing capacity except for two siblings with mild asthma whose values reverted to the normal range after administration of inhaled bronchodilators [9]. Likewise, 60 Pi*ZZ and 19 Pi*SZ subjects identified by neonatal screening in the early 1970s in the Sweden screening had a normal lung function at the age of 30 years, apart from 4 Pi*ZZ smokers who had a significantly lower mean transfer coefficient KCO and forced expiratory volume in 1 s (FEV1)/forced vital capacity (FVC) ratio (which may be a sign of incipient emphysema) than the Pi*ZZ never-smokers [10].

However, a study conducted in London in 59 severe AATD pediatric patients, who had serial measurements of functional residual capacity, showed that AATD may be associated with lung function abnormalities even in very young children [11]. In addition, a study of pulmonary function (using whole body plethysmography, spirometry, and diffusing capacity) held in Berlin in 17 Pi*ZZ schoolchildren, aged between 7 and 18 years, confirmed the absence of pulmonary function abnormalities in the vast majority of subjects, except for an 18-year-old Pi*ZZ girl with a recent onset of exertional dyspnoea and mild obstructive airway disease [12].

Of note, exposure to environmental tobacco smoke has been associated with lung function decrements, especially in children with an impaired protease–antiprotease balance. In this regard, the detrimental effect of passive smoking has been confirmed in a cross-sectional study of 997 primary and secondary schoolchildren aged 11–13 years, and categorized by Pi phenotype as either Pi*M homozygotes or Pi* heterozygotes (mostly MS phenotypes) [13]. Besides, in a large random sample of 3526 schoolchildren, aged 9–11 years, studied according to the International Study of Asthma and Allergies in Childhood (ISAAC) phase II protocol, it was found that those with low levels of AAT were at increased risk of developing pronounced decrements in pulmonary function, particularly if they were exposed to environmental tobacco smoke [14].

In conclusion, despite the lack of uniformity of the available data, all together should support the implementation of appropriate hygienic measures to avoid exposure of children to parental smoking during their growth, as well as the onset of smoking in adolescence. In this regard, it has been found that among adolescents diagnosed with AATD by neonatal screening, the percentage of smokers is very low and that during his childhood parental smoking is reduced [15].

Although an isolated case has been reported in the literature regarding a teenager with AATD and emphysema treated with off-label augmentation therapy, as last resort with good result [16], this therapy is not recommended in this age group by the ATS/ERS standards [2]

11.2.2 Bronchial Asthma

The relationship between AATD and childhood asthma is a controversial subject. For example, a cross-sectional controlled study in 114 AATD patients found that AATD subjects did not exhibit a greater prevalence of airway hyperresponsiveness as compared with control subjects, but hyperresponsive subjects showed an inverse correlation between AAT blood levels and airway hyperresponsiveness (i.e., the lower the AAT levels, the higher the severity of airway hyperresponsiveness) [17].

In addition, in a random sample of 5629 children, aged 9–11 years, studied according to the ISAAC phase II protocol, MS and MZ heterozygosity was not associated with an increased risk of developing asthma, although asthmatics with low serum levels of AAT were particularly prone to develop airway hyperresponsiveness [18].

On the contrary, in the multiple periodic reports of the population-based study from the aforementioned Swedish cohort [4,5,19–23], in general no significant differences in lung function or symptoms were found between AATD and control subjects, but smoking frequency was significantly lower among the AATD subjects than in the controls at age 34 years [5].

11.3 LUNG DISEASE IN ADULTS

11.3.1 COPD

11.3.1.1 Clinical Characteristics

COPD in general, and more specifically pulmonary emphysema of the panacinar type, is the most prevalent clinical manifestation of the AATD and the major cause of disability and death among AATD subjects [2].

Most of the AATD-related COPD patients are Caucasians from northern, western, and central Europe, or white European descendants living in other countries (mainly America, Australia, and New Zealand) [24].

Severe AATD is behind a currently undetermined percentage of COPD cases. Some pioneering studies found between 1.9% and 17.8% of homozygous Pi*Z individuals in referral groups of patients with COPD [25,26], but several subsequent case-finding studies among COPD patients found a mean prevalence of 3.7% (range: 0.4%–10%) for the ZZ genotype, and 1.7% (range: 0.1%–11%) for the SZ genotype [27–36] (Table 11.1).

TABLE 11.1 Results of Several Alpha-1 Antitrypsin Deficiency Case-Finding Studies in Patients With COPD

First author, year, country [reference]	No. of COPD patients	Pi* genotypes				
		ZZ (%)	SZ (%)	MZ (%)	SS (%)	Rare and null (%)
Lieberman, 1986, USA [26]	965	2	0.3	8	0.3	—
Luisetti, 1999, Italy [27]	1841	6	0.9	—	—	0.8
Brantly, 2003, USA [28]	969	3	0.4	11	—	
de la Roza, 2005, Spain [29]	2137	0.4	0.1	—	0.1	
Bals, 2007, Germany [30]	2696	10	2	18	—	0.5
Sitkauskiene, 2008, Lithuany [31]	1167	0.7	0.3	3.4	0.1	—
Greulich, 2012, Germany [32]	11,264	7	1	18	0.3	—
Rahaghi, 2012, USA [33]	3152	0.6	11	—	—	—
Fee, 2015, Ireland [34]	13,500	2	1	14	0.5	—
Sanders, 2015, USA [35]	7530	0.6	0.7	7	0.4	—
Greulich, 2015, Northern and Eastern Europe [36]	1404	2	0.6	11	0.1	—

COPD, Chronic obstructive pulmonary disease.

Pi*MZ heterozygotes are as a rule not considered to be at increased risk for emphysema; however, the results of a metaanalysis were consistent with a small increase risk of COPD in a subpopulation of MZ heterozygotes [37]. In addition, two large population independent studies [i.e., a case-control study from Norway (n = 1669) and a multicenter family-based study from Europe and North America (n = 2707)] found that compared with MM individuals, MZ heterozygotes may be slightly more susceptible to the development of airflow obstruction than MM individuals [38].

In general, individuals with Pi*SZ genotype have not a high risk for lung disease; however, the 10%–20% of individuals with this genotype whose serum AAT levels are below the protective threshold value of 11 μmol/L (empirically assigned to provide clinical guidance in defining pulmonary risk) are at increased risk for lung disease, especially if they smoke [1,39].

Homozygotes with "rare" SERPINA1 alleles associated with low AAT serum levels (e.g., Mmalton), as well as "null" homozygotes, "Z-null," "Z-rare,"

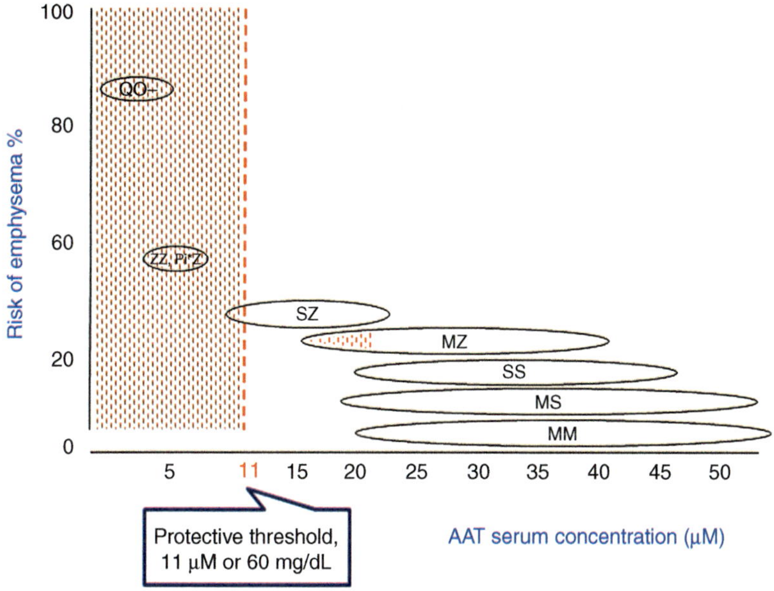

FIGURE 11.1 **Serum concentrations of alpha-1 antitrypsin and risk for COPD.** SERPINA1 genotypes associated with serum AAT levels below the empirically assigned protective threshold value of 11 μmol/L or 60 mg/dL (e.g., QO, ZZ, and 10%–20% of SZ) are considered to confer an increased risk of developing lung disease. For unknown reasons, a small subset of ZZ people with AAT levels above this protection cutoff has increased risk of developing COPD. In any case there is great variability in the clinical expression of the AATD, and environmental inflammatory insults and genetic modifiers that remain poorly understood can accelerate lung destruction.

or "null-rare" compound heterozygotes, are at high risk of developing lung disease [1] (Fig. 11.1).

For its wide social acceptance and the lack of conviction about its harmful effects among many consumers, smoking is the most important risk factor for the development of emphysema in AATD. The decline of FEV1 and mortality is directly influenced by the amount of cigarettes (in terms of packyears) consumed, and patients who quit smoking may reduce their FEV1 decline, but it does not reach the normal values found in nondeficient individuals [40,41].

Paradoxically, some nonsmokers and no-proband Pi*ZZ individuals, who have no other risk factors, may have a life expectancy similar to that of the general population, but other without smoking can also develop lung or liver disease [2].

In smokers with severe AATD symptoms can appear at 35–45 years of age, while among nonsmokers they usually manifest one or two decades later [1–3].

A good example of the negative effect of smoking on the lung of AATD individuals has been provided by a prospective pioneering work by Junus et al., who found that among 33 patients with emphysema the mean age of onset of dyspnoea in smokers (32 years) was significantly lower than that in nonsmokers (51 years); the mean age at death was 48 years in smokers versus 67 in nonsmokers; in nonsmokers the mean FEV1 was 77% versus only 38% in smokers; in smokers, the FEV1 decline correlated with the extent of cigarette smoking pack-years; mean decrease in FEV1 in nonsmokers was abnormally high (80 mL/year), but significantly less than the excessive decline of 317 mL/year in smokers; in ex-smokers the mean rate of decrease was 61 mL/year [42]. Similar findings, summarized in the American Thoracic Society/European Respiratory Society Statement [2], have been reported by other authors.

Besides active cigarette smoking, some environmental exposures (such as indoor kerosene heating, environmental pollutants used in agriculture, mineral dust, gas and fumes, and passive smoking), as well as a history of wheezing, recurrent respiratory infections and pneumonia, can accelerate the development of emphysema in male subjects over 50 years of age [43–46].

11.3.1.2 Pathology

At autopsy, panacinar emphysema with basal predominance is usually seen in patients with severe AATD. On occasion, large bullae, preferentially in the basal parts of the lungs, are described at autopsy or in surgical specimens. Occasionally, minimal centrilobular emphysema is observed in the upper lobes [2].

In 280 atypical lung resection specimens of 81 German patients, various forms of emphysema were found; in 66% there were bullous changes focally; acute bronchiolitis was seen in 42%, slight chronic bronchiolitis in 27%, and bronchioloectasia in 45% of the specimens. Focal intraalveolar aggregates of granulocytes were identified in 16% and "organized pneumonia" in 21%. Occult neoplasms were found in 10% of the patients [47].

In addition, the histopathologic specimens of lungs and liver of 42 individuals who died while enrolled in the American AATD Registry were reviewed. The underlying cause of death was emphysema in 34 patients and cirrhosis in 2 patients. All lungs exhibited severe panacinar emphysema with a lower lobe predominance. Centriacinar emphysema was minimal. There were minimal changes in small airways, and mildly increased bronchial gland-to-wall ratio with goblet cell hyperplasia. Dilatation of membranous bronchioles was a frequent finding; however, bronchiectasis of larger airways was a minor feature in only 6 patients (15%) [48].

A distinctive histopathological finding of individuals with Pi*ZZ AATD-related emphysema is that polymeric Z-AAT colocalizes with neutrophils in the lung alveoli [49].

More recently, an important adaptive immune inflammation, comprising B, CD4(+) and CD8(+) lymphocytes, and lymphoid follicles, have been demonstrated in explanted lungs from both subjects with AATD and subjects with usual COPD [50].

11.3.1.3 Symptoms

Symptoms are similar to those experienced by patients with COPD of any etiology [2].

Dyspnea is the most common symptom, and is present in 80%–90% of individuals. It is initially evident only during physical exertion, but it progresses over time limiting the patient's capacity to develop light physical activity, and eventually, in extreme cases, it can occur even at rest. Between 65% and 75% of patients refer wheezing either routinely or in relation to respiratory infections, and up to 40% complain of nonproductive or productive cough; the latter sometimes related to the presence of bronchiectasis [51,52].

11.3.1.4 Physical Examination

Patients with mild emphysema generally have no abnormal findings on physical examination. In mild or moderate stages, there are no specific findings for AATD, and disease detection is easily missed if the physician relies solely on physical findings. Advanced pulmonary disease is associated with nonspecific sings of chest hyperinflation, including barrel chest deformation and increased percussion note. Increased respiratory work is evident as tachypnea, tachycardia, scalene and intercostal muscle retraction, and tripod position. Airflow obstruction manifests as pursed-lip breathing, wheezing, and paradoxical pulse. On auscultation it is common to find a decreased breath sounds intensity and distant heart sounds (both more marked, the greater the severity of emphysema is); crackles at the bases and diffuse wheezing can be detected especially during exacerbations [2].

With disease progression weight loss and muscle wasting (especially atrophy of the muscles of the lower extremities) may occur. At advanced stages many patients are underweight due to the systemic effect of chronic lung inflammation. Cyanosis and signs of right heart failure or cor pulmonale are usually present only in very advanced stages of the disease once respiratory failure and pulmonary hypertension occur [2].

11.3.1.5 Lung Function Test

Spirometry is the basic lung function test for its accessibility and reproducibility. It typically reveals bronchial obstruction, with reduced FEV1, normal or reduced FVC, and FEV1/FVC <70% ratio. In emphysema, bronchial obstruction is due to the loss of elasticity of the lung parenchyma causing dynamic airway collapse. The loss of elastic recoil leads to air trapping with hyperinflation resulting in increased total lung capacity and residual volume. The flow volume curve shows a reduction in lung flows, with a concave morphology [2]. The severity of spirometry impairment relates to the severity of the disease and patients' survival [53]. Of note, 17% of smokers with spirometric parameters within normal limits and without clinical symptoms had emphysema on CT scan [54].

Pulmonary parenchyma destruction determines a loss of gas exchange surface which results in an increase of the alveolar–arterial oxygen gradient. The small pulmonary vessels destruction reduces the Carbon Monoxide Diffusing Capacity, and this parameter fairly accurately correlates with the loss of lung density measured with chest CT scan as well as with the anatomical degree of emphysema [55,56].

The progressive loss of lung function can lead to respiratory failure. Initially, this is manifested by hypoxemia (i.e., reduction in arterial oxygen pressure, PaO_2) during exercise. In advanced stages, the PaO_2 may be reduced at rest. Hypercapnia (i.e., retention of carbon dioxide, CO_2, in arterial blood) is a rare event which only appears in final stages [2].

Exercise tolerance may be limited due to a reduced ventilatory capacity and dynamic lung hyperinflation, and in severe cases these pathophysiological events may cause intolerable dyspnea during minor physical activities [57]. The 6-min walk test (6MWT) is used in clinical settings to evaluate exercise capacity at submaximal exercise levels in COPD [58], and it is known that the exercise capacity and the degree of desaturation in the 6MWT are two important factors that correlate quite well with the health status in AATD–COPD patients [59].

11.3.1.6 Radiology

The chest radiograph is usually normal in early stages. In more advanced stages, radiological signs of air trapping (i.e., flattening of the diaphragm, increased retrosternal space, and horizontally aligned ribs) along with other signs of lung tissue destruction (i.e., hyperlucency, vascular attenuation -or oligemia- and bullae) may be observed. (Fig. 11.1). Although these radiologic data are not specific, in the AATD-associated emphysema, lung hyperlucency and bullae predominate in lower lobes, as it is typical of panacinar emphysema, while in other types of emphysema (including centriacinar) mainly affect the upper lobes [60].

The high-resolution chest computed tomography (HRCT) scan allows detecting early emphysematous changes and bronchiectasis, and currently it is accepted by most experts that HRCT is clearly more sensitive to detect the presence of emphysema than the simple chest X-ray, and even more than the respiratory function tests [61]. As said before, emphysema may be present on chest CT without evidence of impaired pulmonary function tests [54]. The HRCT easily shows areas of low density (or hyperlucency), corresponding to lung areas of parenchymal destruction with reduced pulmonary vasculature. It can also detect very reliably bronchial wall thickening and/or dilatation with bronchiectasis. Bullae formation usually occurs in advanced stages of the disease. The process is not uniform, and some areas are more affected than other. Estimation of the extent of disease can correlate well with lung function tests. In contrast to the usual pattern observed in the centriacinar emphysema (i.e., emphysematous changes more pronounced in the lung apices than bases), emphysematous changes observed in two-thirds of individuals with AATD are more pronounced in bases than apices [62,63] (Figs. 11.2 and 11.3).

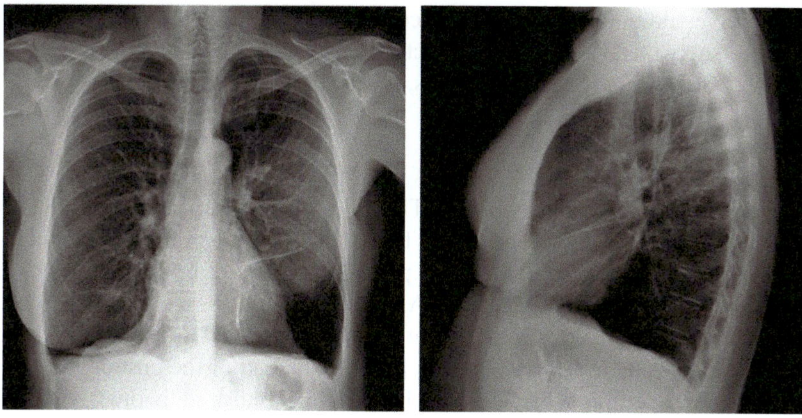

FIGURE 11.2 **Simple chest radiographs of a female with ZZ alpha-1 antitrypsin deficiency-related pulmonary emphysema.** Figure adapted from the courtesy of Dr. Francisco Casas (Granada, Spain). Posteroanterior and profile projection chest radiograph, showing hyperinflation with basal hyperlucency of both lungs, albeit hyperlucency is more pronounced in the left base for a mastectomy that the patient underwent years before.

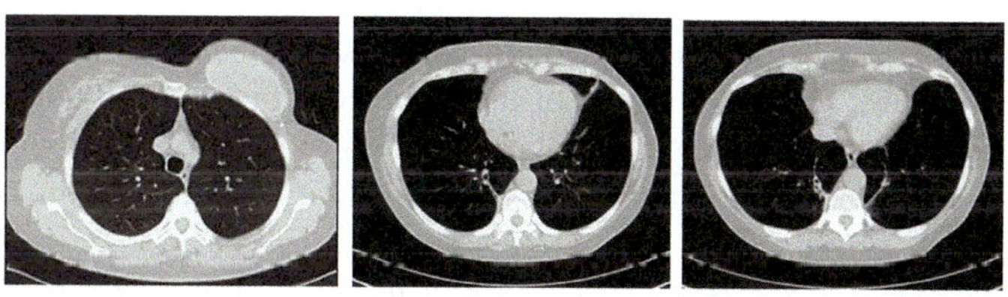

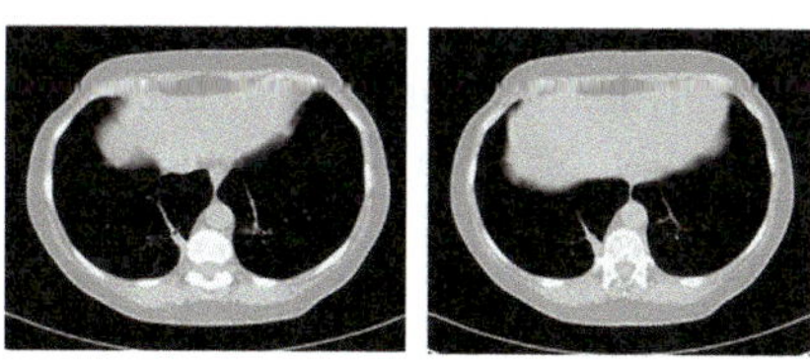

	Lung function parameters					
	FVC % pred	FEV_1 % pred	FEV_1 % FVC	TLC % pred	RV % pred	DLCO % pred
Basal	75.2	59.6	67.1	117.1	180.9	49.6
Postbronchodilator %	80.4 (+7)	65 (+9)	68.4 (+19)	–	–	–

FIGURE 11.3 **High-resolution computed tomography of the chest and lung function test results of the same ZZ female with alpha-1 antitrypsin deficiency-related pulmonary emphysema.** Figure adapted from the courtesy of Dr. Francisco Casas (Granada, Spain). *FEV1*, Forced expiratory volume in 1 s; *FVC*, forced vital capacity; *DLCO*, Carbon Monoxide Diffusing Capacity; *RV*, residual volume; *TLC*, total lung capacity. The basal cuts of the CT scan show more pronounced emphysematous changes than the more apical section (first upper left image) of the lung. Pulmonary function reveals a slight obstructive pattern with air trapping, nonsignificant postbronchodilator reversibility, and moderate reduction in diffusion capacity.

Of note, although it is generally accepted that emphysematous changes related to severe AATD primarily affect lower lobes, some studies have found greater involvement in the upper lobes, especially in Pi*SZ patients [64].

11.3.2 Bronchiectasis

The causal relationship between AATD and bronchiectasis is a controversial subject because although some studies have found a high prevalence of bronchiectasis in AATD patients, the general population series of bronchiectasis have not shown an increased frequency of AATD cases.

Historically, two of the first five patients diagnosed by Laurell and Eriksson in the early 1960s had bronchiectasis [65]. At the end of the ensuing decade, Larsson described bronchiectasis in 11% of 246 Swedish individuals with ZZ phenotype [40]. Other authors at different times reported isolated cases or small series of Pi*Z patients with cystic or varicose bronchiectasis at times without or with minimal emphysema [66–69], and the NHLBI registry reported bronchiectasis in only 2% of 1129 registered participants [70].

The results of a case control study in 202 patients with bronchiectasis did not support a physiopathologic implication of the AAT genes in the development of bronchiectasis, suggesting that they may be a consequence of emphysema in Pi*Z patients rather than a primary effect [71].

A more detailed study in a group of 74 Pi*Z subjects found bronchiectatic changes on chest CT in 70 (95%) of these subjects, but clinically significant bronchiectasis (i.e., radiologic bronchiectasis in 4 or more bronchopulmonary segments together with symptoms of regular sputum production) in only 20 (27%) [72].

In addition, a recent descriptive study of the Spanish Registry of Patients with AATD reported bronchiectasis in 32% of ZZ, 18% of SZ, and 24% of patients with rare genotypes (the majority Mmalton) [73].

In summary, the available data are currently insufficient to draw conclusions about a possible, but not yet proven, association between AATD and bronchiectasis.

11.3.3 Asthma

Asthma and AATD are two different conditions yet they can coexist, and in clinical practice, some patients with AATD initially present with asthma-like symptoms and receive treatment for asthma.

A study from the NHLBI American Registry on 1129 subjects, 18 years of age or older with severe AATD, found that 45% of enrollees reported a history of asthma-like symptoms, and more than 61% showed a positive bronchodilator test [51]. A later study from the NHLBI American Registry, assessing the presence of wheezing, atopy, elevated serum IgE, and positive bronchodilator response as markers of asthma, found three or more of these markers in 22% of individuals with AATD, and in only 5% of those with COPD without AATD, suggesting that a lack of AAT in airways increases the propensity to develop asthma [52].

However, in a prospective study in 1060 German patients with COPD and asthma, screened for AATD using dried blood spots, found that the prevalence of AATD did not differ from that found in the general population [74]. In addition, a cross-sectional study on a sample of 111 adult patients with asthma found that the distribution of AAT phenotypes in asthmatic patients did not differ from that found in the general population, and that heterozygote phenotypes did not appear to confer greater severity or different clinical expression [75]. Finally, a further study conducted in the Netherlands in 122 patients with severe asthma concluded that heterozygosity of AAT does not seem to be an important risk factor of persistent airflow limitation in patients with asthma [76].

In summary, the available data are currently insufficient to draw conclusions about the association between AATD and bronchial asthma, and longitudinal follow-up well-designed studies with larger sample sizes are needed to obtain reliable conclusions.

REFERENCES

[1] Stoller JK, Lacbawan FL, Aboussouan LS. Alpha-1 antitrypsin deficiency. In: Pagon RA, Adam MP, Ardinger HH, et al., editors, GeneReviews® [Internet]. Seattle, WA: University of Washington; 1993–2016. Available from: http://www.ncbi.nlm.nih.gov/books/NBK1519/

[2] American Thoracic Society/European Respiratory Society. American Thoracic Society/European Respiratory Society statement: standards for the diagnosis and management of individuals with alpha1-antitrypsin deficiency. Am J Resp Clin Care Med 2003;168:818–900.

[3] Silverman EK, Pierce JA, Province MA, Rao DC, Campbell EJ. Variability of pulmonary function in alpha-1-antitrypsin deficiency: clinical correlates. Ann Intern Med 1989;111:982–91.

[4] Sveger T, Piitulainen E, Arborelius M Jr. Lung function in adolescents with alpha 1-antitrypsin deficiency. Acta Paediatr 1994;83:1170–3.

[5] Tanash HA, Nystedt-Düzakin M, Cano Montero L, Sveger T, Piitulainen E. The Swedish alpha 1-antitrypsin screening study: health status, lung and liver function at age 34. Ann Am Thorac Soc 2015;12:807–12.

[6] Talamo RC. Basic and clinical aspects of the alpha1-antitrypsin. Pediatrics 1975;56:91–9.

[7] Cox DW, Talamo RC. Genetic aspects of pediatric lung disease. Pediatr Clin North Am 1979;26:467–80.

[8] Wagener JS, Sobonya RE, Taussig LM, Lemen RJ. Unusual abnormalities in adolescent siblings with alpha 1-antitrypsin deficiency. Chest 1983;83:464–8.

[9] Wall M, Moe E, Eisenberg J, Powers M, Buist N, Buist AS. Long-term follow-up of a cohort of children with alpha-1-antitrypsin deficiency. J Pediatr 1990;116:248–51.

[10] Bernspång E, Wollmer P, Sveger T, Piitulainen E. Lung function in 30-year-old alpha-1-antitrypsin-deficient individuals. Respir Med 2009;103:861–5.

[11] Hird MF, Greenough A, Mieli-Vergani G, Mowat AP. Hyperinflation in children with liver disease due to alpha-1-antitrypsin deficiency. Pediatr Pulmonol 1991;11:212–6.

[12] Wiebicke W, Niggemann B, Fischer A. Pulmonary function in children with homozygous alpha1-protease inhibitor deficiency. Eur J Pediatr 1996;155:603–7.

[13] Corbo GM, Forastiere F, Agabiti N, Dell'Orco V, Pistelli R, Massi G, et al. Passive smoking and lung function in α1-antitrypsin heterozygote children. Thorax 2003;58:237–41.

[14] von Ehrenstein OS, von Mutius E, Maier E, Hirsch T, Carr D, Schaal W, et al. Lung function of school children with low levels of α1-antitrypsin and tobacco smoke exposure. Eur Respir J 2002;19:1099–106.

[15] Thelin T, Sveger T, McNeil TF. Primary prevention in a high-risk group: smoking habits in adolescents with homozygous alpha-1-antitrypsin deficiency (ATD). Acta Paediatr 1996;85:1207–12.

[16] Griese M, Brüggen H. Intravenous alpha1-antitrypsin in a child with deficiency and severe lung disease. Eur Respir J 2009;34:278–9.

[17] Malerba M, Radaeli A, Ceriani L, Tantucci C, Grassi V. Airway hyperresponsiveness in a large group of subjects with alpha1-antitrypsin deficiency: a cross-sectional controlled study. J Intern Med 2003;253:351–8.

[18] von Ehrenstein OS, Maier EM, Weiland SK, Carr D, Hirsch T, Nicolai T, et al. Alpha1 antitrypsin and the prevalence and severity of asthma. Arch Dis Child 2004;89:230–1.

[19] Piitulainen E, Tornling G, Eriksson S. Effect of age and occupational exposure to airway irritants on lung function in non-smoking individuals with alpha-1-antitrypsin deficiency (PiZZ). Thorax 1997;52:244–8.

[20] Piitulainen E, Tornling G, Eriksson S. Environmental correlates of impaired lung function in non-smokers with severe alpha-1-antitrypsin deficiency (PiZZ). Thorax 1998;53:939–43.

[21] Piitulainen E, Sveger T. Effect of environmental and clinical factors on lung function and respiratory symptoms in adolescents with alpha-1 antitrypsin deficiency. Acta Paediatr 1998;87:1120–4.

[22] Sveger T. Prospective study of children with alpha 1-antitrypsin deficiency: eight-year-old follow-up. J Pediatr 1984;104:91–4.

[23] Piitulainen E, Sveger T. Respiratory symptoms and lung function in young adults with severe alpha(1)-antitrypsin deficiency (PiZZ). Thorax 2002;57:705–8.

[24] de Serres FJ, Blanco I, Fernández-Bustillo E. Estimating the risk for alpha-1 antitrypsin deficiency among COPD patients: evidence supporting targeted screening. COPD 2006;3:133–9.

[25] Cox DW, Hoeppner VH, Levison H. Protease inhibitors in patients with chronic obstructive pulmonary disease: the alpha1-antitrypsin heterozygote controversy. Am Rev Respir Dis 1976;113:601–6.

[26] Lieberman J, Winter JB, Sastre A. Alpha-1-antitrypsin PI-types in 965 COPD patients. Chest 1986;89:370–3.

[27] Luisetti M, Massi G, Massobrio M, Guarraci P, Menchicchi FM, Beccaria M, et al. A national program for detection of alpha 1-antitrypsin deficiency in Italy. Gruppo I. D. A.. Respir Med. 1999;93:169–72.

[28] Brantly ML, Mishra VS, Viranovskaya N, Zienko L, Corcoran V, Leonard S. Statewide targeted screening and detection of AAT deficiency. Am J Respir Crit Care Med 2003;167:A222.

[29] de la Roza C, Rodríguez-Frías F, Lara B, Vidal R, Jardí R, Miravitlles M. Results of a case-detection programme for alpha1-antitrypsin deficiency in COPD patients. Eur Respir J 2005;26:616–22.

[30] Bals R, Koczulla R, Kotke V, Andress J, Blackert K, Vogelmeier C. Identification of individuals with alpha-1-antitrypsin deficiency by a targeted screening program. Respir Med 2007;101:1708–14.

[31] Sitkauskiene B, Serapinas D, Blanco I, Fernández-Bustillo E, Janciauskiene S, Sakalauskas R. Screening for alpha1-antitrypsin deficiency in Lithuanian patients with COPD. Respir Med 2008;102:1654–8.

[32] Greulich T, Nell C, Kehr K, Kotke V, Wiedmann S. Detection of patients with alpha-1 antitrypsin deficiency in Germany—update 2011. In: COPD: alpha 1 antitrypsin (American Thoracic Society International Conference Abstracts, 2012, vol. 183). American Thoracic Society: abstract A4356.

[33] Rahaghi F, Ortega I, Rahaghi N, Oliveira E, Ramirez J, Smolley L, et al. Physician alert suggesting alpha-1 antitrypsin deficiency testing in pulmonary function test (PFT) results. COPD 2009;6:26–30.

[34] Fee L, Carroll T, O'Connor C, Ryan P, O'Brien P, Pentony E. The Irish Alpha-1 Antitrypsin Deficiency National Targeted Detection Programme. C46. How do you do it? COPD Diagnosis Assess 2015;191:A4477.

[35] Sanders C, Kim J. Frequencies of alpha-1 antitrypsin deficiency phenotypes detected using a clinical testing strategy that reflexes to next generation sequencing. In: C45. Across the Universe of COPD Epidemiology (American Thoracic Society International Conference Abstracts, 2015, vol. 191). American Thoracic Society: abstract A4456.

[36] Greulich T, Averyanov A, Borsa L, Rozborilová E, Vaicius D, Major T, et al. European screening for alpha-1 antitrypsin deficiency in subjects with lung disease. Clin Respir J 2015; [Epub ahead of print].

[37] Hersh CP, Dahl M, Ly NP, Berkey CS, Nordestgaard BG, Silverman EK. Chronic obstructive pulmonary disease in alpha1-antitrypsin PI MZ heterozygotes: a meta-analysis. Thorax 2004;59:843–9.

[38] Sørheim IC, Bakke P, Gulsvik A, Pillai SG, Johannessen A, Gaarder PI, et al. α1-Antitrypsin protease inhibitor MZ heterozygosity is associated with airflow obstruction in two large cohorts. Chest 2010;138:1125–32.

[39] Turino GM, Barker AF, Brantly ML, Cohen AB, Connelly RP, Crystal RG, et al. Clinical features of individuals with PI*SZ phenotype of alpha-1 antitrypsin deficiency. Alpha-1 Antitrypsin Deficiency Registry Study Group. Am J Respir Crit Care Med 1996;154:1718–25.

[40] Larsson C. Natural history and life expectancy in severe alpha-1 antitrypsin deficiency PiZ. Acta Med Scand 1978;294:345–51.

[41] Wu MC, Eriksson S. Lung function, smoking and survival in severe alpha 1-antitrypsin deficiency, PiZZ. J Clin Epidemiol 1988;41:1157–65.

[42] Janus ED, Phillips NT, Carrell RW. Smoking, lung function, and alpha 1-antitrypsin deficiency. Lancet 1985;1:152–4.

[43] Buist AS, Burrows B, Eriksson S, Mittman C, Wu M. The natural history of air-flow obstruction in PiZ emphysema. Am Rev Respir Dis 1982;127:S43–50.

[44] Hutchinson DCS. Natural history of alpha-1-protease inhibitory deficiency. Am J Med 1988;84(Suppl. 6A):3–12.

[45] Seersholm N, Kok-Jensen A, Dirksen A. Decline in FEV1 among patients with severe hereditary alpha-1 antitrypsin deficiency type PiZ. Am J Respir Crit Care Med 1995;152:1922–5.

[46] Dowson LJ, Guest PJ, Stockley RA. Longitudinal changes in physiological, radiological and health status measurements in alpha(1)antitrypsin deficiency and factors associated with decline. Am J Respir Crit Care Med 2001;164:1805–9.

[47] Theegarten D, Teschler H, Stamatis G, Konietzko N, Morgenroth K. Pathologico-anatomical results in surgical lung volume reduction of advanced emphysema. Pneumologie 1998;52:702–6.

[48] Tomashefski JF Jr, Crystal RG, Wiedemann HP, Mascha E, Stoller JK. Alpha 1-Antitrypsin Deficiency Registry Study Group. The bronchopulmonary pathology of alpha-1 antitrypsin (AAT) deficiency: findings of the Death Review Committee of the national registry for individuals with Severe Deficiency of Alpha-1 Antitrypsin. Hum Pathol. 2004;35:1452–61.

[49] Mahadeva R, Atkinson C, Li Z, Stewart S, Janciauskiene S, Kelley DG, et al. Polymers of Z alpha1-antitrypsin co-localize with neutrophils in emphysematous alveoli and are chemotactic in vivo. Am J Pathol 2005;166:377–86.

[50] Baraldo S, Turato G, Lunardi F, Bazzan E, Schiavon M, Ferrarotti I, et al. Immune activation in α1-antitrypsin-deficiency emphysema. Beyond the protease-antiprotease paradigm. Am J Respir Crit Care Med 2015;191:402–9.

[51] McElvaney NG, Stoller JK, Buist AS, Prakash UB, Brantly ML, Schluchter MD, et al. Baseline characteristics of enrolees in the National Heart, Lung and Blood Institute Registry of alpha-1 antitrypsin deficiency. Alpha-1Antitrypsin Deficiency Registry Study Group. Chest 1997;111:394–403.

[52] Eden E, Mitchell D, Mehlman B, Khouli H, Nejat M, Grieco MH, et al. Atopy, asthma, and emphysema in patients with severe alpha-1-antitrypysin deficiency. Am J Respir Crit Care Med 1997;156:68–74.

[53] Seersholm N, Dirksen A, Kok-Jensen A. Airways obstruction and two year survival in patients with severe alpha 1-antitrypsin deficiency. Eur Respir J 1994;7:1985–7.

[54] Smith BM, Austin JH, Newell JD Jr, D'Souza BM, Rozenshtein A, Hoffman EA, et al. Pulmonary emphysema subtypes on computed tomography: the MESA COPD study. Am J Med 2014;127:94.e7–94.e23.

[55] Morrison NJ, Abboud RT, Ramadan F, Miller RR, Gibson NN, Evans KG, et al. Comparison of single breath carbon monoxide diffusing capacity and pressure-volume curves in detecting emphysema. Am Rev Respir Dis 1989;139:1179–87.

[56] Gould GA, Redpath AT, Ryan M, Warren PM, Best JJ, Flenley DC, et al. Lung CT density correlates with measurements of airflow limitation and the diffusing capacity. Eur Respir J 1991;4:141–6.

[57] Task Force ERS, Palange P, Ward SA, Carlsen KH, Casaburi R, Gallagher CG, et al. Recommendations on the use of exercise testing in clinical practice. Eur Res J 2007;29:185–209.

[58] Butland RJ, Pang J, Gross ER, Woodcock AA, Geddes DM. Two-, six-, and 12-minute walking tests in respiratory disease. Br Med J (Clin Res Ed) 1982;284:1607–8.

[59] Dowson LJ, Newall C, Guest PJ, Hill SL, Stockley RA. Exercise capacity predicts health status in alpha(1)-antitrypsin deficiency. Am J Respir Crit Care Med 2001;163:936–41.

[60] Gishen P, Saunders AJS, Tobin MJ, Hutchison DCS. Alpha-1 antitrypsin deficiency: the radiological features of pulmonary emphysema in subjects of Pi-type-Z and Pi-type-SZ. A survey by the British Thoracic Association. Clin Radiol 1982;33:371–7.

[61] Parr DG, Stoel BC, Stolk J, Stockley RA. Pattern of emphysema distribution in alpha-1 antitrypsin deficiency influences lung function impairment. Am J Respir Crit Care Med 2004;170:1172–8.

[62] Guest PJ, Hansell DM. High resolution computed tomography in emphysema associated with alpha-1-antritrypsin deficiency. Clin Radiol 1992;45:260–6.

[63] Wilson JS, Galvin JR. Normal diffusing capacity in patients with PiZ alpha-1-antitrypsin deficiency, severe airflow obstruction and significant radiographic emphysema. Chest 2000;118:867–71.

[64] Holme J, Stockley RA. CT scan appearance, densitometry, and health status in protease inhibitor SZ alpha1-antitrypsin deficiency. Chest 2009;136:1284–90.

[65] Laurell CB, Eriksson S. The electrophoretic alpha1-globulin pattern of serum in alpha1-antitrypsin deficiency. Scand J Clin Lab Invest 1963;15:132–40.

[66] Erikson S. Studies in alpha 1 antitrypsin deficiency. Acta Med Scand Suppl 1965;432:1–85.

[67] Blanco I, Menéndez ML, Carro F. Extensive bronchiectasis as unique manifestation of a severe case of alpha 1 antitrypsin deficiency. Arch Bronconeumol 1992;28:253.

[68] Blanco I, Canto H, Carro F. Bronchiectasis in patients with alpha 1 antitrypsin severe deficiency: a frecuent association? Arch Bronconeumol 1994;29:473.
[69] King MA, Stone JA, Diaz PT, Mueller CF, Becker WJ, Gadek JE. Alpha1 antitrypsin deficiency: evaluation of bronchiectasis with CT. Radiology 1996;199:137–41.
[70] Fallat RJ. Reactive airways disease and alpha 1-antitrypsin deficiency. In: Crystal RG, editor. Alpha 1-Antitrypsin Deficiency: Biology, Pathogenesis, Clinical Manifestations, Therapy. New York, NY: Marcel Dekker; 1996. p. 259–79.
[71] Cuvelier A, Muir JF, Hellot MF, Benhamou D, Martin JP, Bénichou J, et al. Distribution of alpha(1)-antitrypsin alleles in patients with bronchiectasis. Chest 2000;117:415–9.
[72] Parr DG, Guest PG, Reynolds JH, Dowson LJ, Stockley RA. Prevalence and impact of bronchiectasis in alpha-1antitrypsin deficiency. Am J Respir Crit Care Med 2007;176:1215–21.
[73] Lara B, Miravitlles M. Spanish registry of patients with alpha-1 antitrypsin deficiency; comparison of the characteristics of PISZ and PIZZ individuals. COPD 2015;12(Suppl. 1):27–31.
[74] Wencker M, Marx A, Konietzko N, Schaefer B, Campbell EJ. Screening for alpha-1 Pi deficiency in patients with lung diseases. Eur Respir J 2002;20:319–24.
[75] Miravitlles M, Vilà S, Torrella M, Balcells E, Rodríguez-Frías F, de la Roza C, et al. Influence of deficient alpha1-anti-trypsin phenotypes on clinical characteristics and severity of asthma in adults. Respir Med 2002;96:186–92.
[76] van Veen IH, ten Brinke A, van der Linden AC, Rabe KF, Bel EH. Deficient alpha-1-antitrypsin phenotypes and persistent airflow limitation in severe asthma. Respir Med 2006;100:1534–9.

Chapter 12

Other Diseases Associated With Alpha-1 Antitrypsin Deficiency

12.1 INTRODUCTION

Apart from the unanimously recognized COPD, intrahepatic cholestasis, and liver cirrhosis, only neutrophilic panniculitis and Wegener's granulomatosis have been formally associated with alpha-1 antitrypsin (AAT) deficiency (AATD).

A variety of other illnesses which have been observed in individuals with AATD [e.g., rheumatoid arthritis, inflammatory bowel disease (IBD), fibromyalgia/chronic fatigue syndrome, arterial aneurysms and dissections, diabetes mellitus, hypothyroidism, several types of cancers, etc.] can be simple random occurrences of various diseases in the same individual with AATD. More studies showing that these relationships are causal and not casual are still needed to accept that they are not just chance occurrences.

12.2 GRANULOMATOSIS WITH POLYANGIITIS

Granulomatosis with polyangiitis (GPA; previously known as "Wegener's granulomatosis") is an idiopathic vasculitis of small and medium-sized blood vessels, characterized by necrotizing granulomatous inflammation. It affects people of both sexes, especially Caucasians with Pi*ZZ genotype, at any age, with predominance of the 6th and 7th decade of life. GPA is a potentially fatal illness which typically affects upper and lower respiratory tracts with coexisting glomerulonephritis. In 80%–95% of the patients, the first symptoms are otorhinolaryngological manifestations of nose/sinuses, ears, eyes, larynx/trachea, oral cavity, or salivary glands. Recurrent sinusitis is usually severe. Lung nodules (some cavitated) and transient infiltrates (by alveolar haemorrhages) are frequently seen. Renal involvement is characterized by rapidly progressive severe necrotizing glomerulonephritis with extracapillary crescents. GPA is usually characterized by antineutrophil cytoplasmic antibodies (ANCAs), although there is less frequently cases with negative ANCA [1]. In clinical practice, the diagnosis is based on a compatible clinical picture and the presence of blood ANCA antibodies with a biopsy-proven vasculitis of affected organs (Fig. 12.1). The standard treatment procedure is based on combination

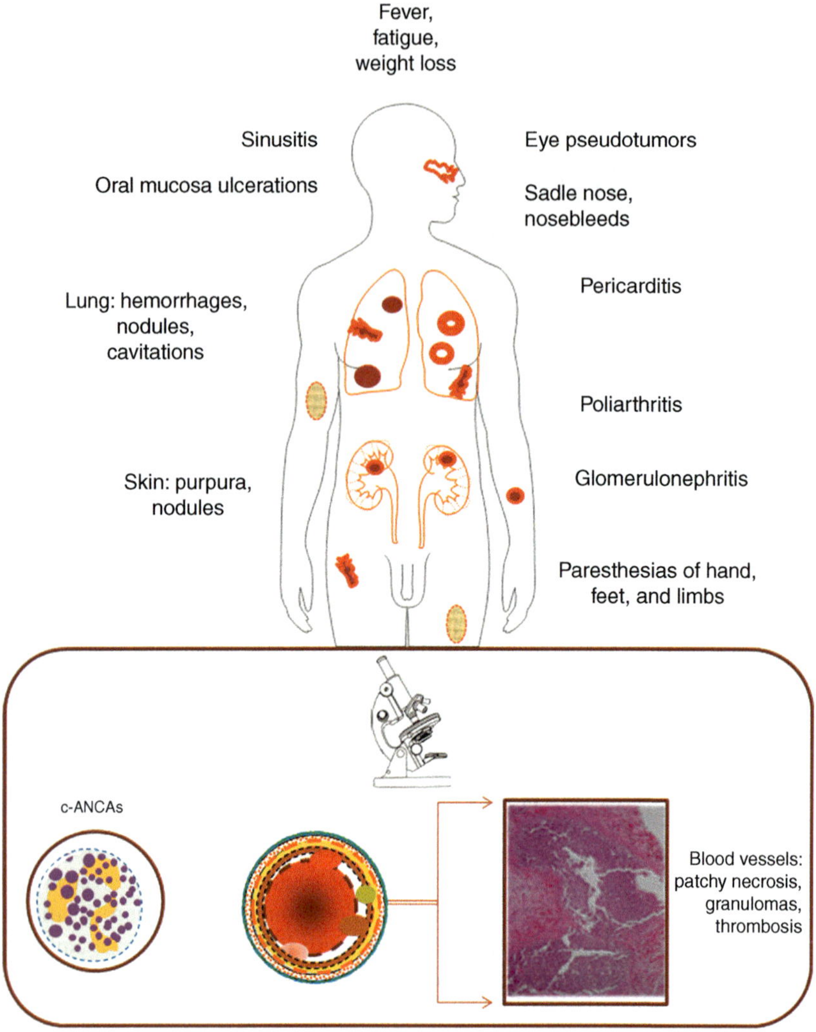

FIGURE 12.1 Signs and symptoms, and laboratory findings of granulomatosis with polyangiitis. Granulomatosis with polyangiitis (also known as "Wegener's granulomatosis") is a potentially fatal systemic necrotizing vasculitis, involving the upper and lower respiratory tract and kidneys, and less frequently other organs, which affects small- and medium-sized blood vessels and is usually associated with c-ANCAs (cytoplasmic antineutrophil cytoplasmic antibodies).

of systemic corticosteroid and immunosuppressant therapy. However, relapses remain frequent (50% at 5 years), and maintenance treatment is now the main therapeutic challenge [2].

ANCAs are a group of autoantibodies, mainly of the IgG type, against antigens located in the cytoplasm of neutrophil granulocytes. They can be classified according to their immunofluorescence staining pattern in several different types, the most typical the cytoplasmic ANCA (c-ANCA) and the perinuclear ANCA (p-ANCA); c-ANCA and p-ANCA antigens are specifically proteinase 3 (PR3) and myeloperoxidase (MPO), respectively. PR3 and MPO are serine proteases packed in the neutrophil azurophilic (or "primary") granules, together to elastase, defensins, cathepsin G, neutral serine proteases, phospholipase A2, acid hydrolases, bactericidal/permeability-increasing protein, lysozyme, and proteoglycans. It is poorly understood how ANCAs are developed, but, among other hypotheses, it has been attributed to a defective neutrophil apoptosis, an essential process for controlling the duration of the early inflammatory response, restricting tissues damage by neutrophil proteases. Of note, in AATD neutrophil apoptosis is significantly increased due to an abnormal activation of the TNF-α system [3], and ANCA may be developed either via ineffective apoptosis or via ineffective removal of apoptotic cell fragments, leading to the exposure of the immune system to molecules normally sequestered inside the cells [4]. PR3-directed c-ANCAs are present in 80%–90% of GPA. These antibodies have a direct pathological role in small vessel vasculitides, activating neutrophils through their Fc surface receptors, causing neutrophils to display PR3 and MPO on their surface. Activated neutrophils can then adhere to endothelial cells, releasing free oxygen radicals, proteolytic enzymes, and chemoattractive signalling molecules that recruit other neutrophils to the endothelium, finally resulting in vascular damage with ischemia of the affected organ [5]. It may be assumed that these pathogenic mechanisms would be more pronounced in AATD people with insufficient production of AAT to inhibit the excess of released inflammatory products.

As of the 1990s several studies have found significantly higher frequencies of the Z allele in patients with GPA compared with controls, all together suggesting that AATD may be a driving factor for the development of this type of vasculitis [6–15].

The results of these studies are in line with those provided by a potent epidemiological study conducted in Boston using samples from 433 patients of the Wegener's granulomatosis Genetic Repository and 576 controls, which found that compared with the normal MM genotype the odds ratio for MZ or MS genotypes was 1.5, and the odds ratio for ZZ, SS, or SZ genotypes was 15, these findings strengthening the evidence of a causal link between AATD and susceptibility to Wegener's granulomatosis development [16].

On the contrary, another powerful study conducted in the United Kingdom showed that the Z allele, but not the S, was significantly elevated in a series of 856 patients with GPA, and in addition the carriers of Z alleles showed higher concentrations of Z-AAT circulating polymers than non-Z. Furthermore, polymer deposits were found in a renal biopsy from an MZ patient with vasculitis-related glomerulonephritis. Of note, such polymers demonstrated proinflammatory activity in vitro [17].

Anecdotally, it has been reported a case of a Pi*ZZ, 49-year-old man with extensive cutaneous leukocytoclastic vasculitis, multiple recurrences, partially controlled with colchicine, prednisone, plasma infusions, plasma-exchange therapy, who 5 years later presented a dramatic response after intravenous treatment with AAT infusions [18]. More recently, it has been reported the case of an 84-year-old Pi*ZZ woman with pulmonary emphysema and multiorgan involvement by GPA, who also experienced a favorable response with AAT infusions [19].

12.3 NEUTROPHILIC PANNICULITIS

Neutrophilic panniculitis associated with AATD is a rare disease, which can be the only manifestation, or accompany or precede the typical respiratory or liver manifestations of AATD [20].

The first case of this disease was diagnosed 45 years ago in the Department of Dermatology, Hospital of the University of Strasbourg, by three French dermatologists (Grosshans, Warter, and Storck) [21,22].

This initial discovery went largely unnoticed until 5 years later, when Rubinstein et al. (Maywood, IL, USA) reported two Pi*ZZ patients with neutrophilic panniculitis and very low serum levels of AAT, hypothesizing that "the marked reduction of antiproteolytic activity found in Pi*ZZ homozygotes may predispose to or aggravate Weber-Christian disease ..." [23].

After these first cases, approximately other 50 cases were reported worldwide, 2/3 of them with ZZ phenotypes [20,24]. Really, the prevalence of neutrophilic panniculitis associated with severe AATD is very low, as evidenced by the fact that only 1 of 1129 participants (i.e., 0.1 %) in the National Heart, Lung, and Blood Institute Registry of Individuals with Severe Deficiency of Alpha-1 Antitrypsin reported had panniculitis [25]; however, more recently the disease was reported by 6 of 651 (0.9%) Pi*ZZ or Pi*Z-null patients from the UK National Centre for AATD [26], and it could be probably higher if AATD were more frequently tested in panniculitis.

The heritage of the AAT gene is not sex-linked, so it is transmitted in similar proportion in men and women, and yet in the AATD-associated panniculitis cases described in the literature there is a clear predominance of women, aged between 30 and 60 years, with Pi*ZZ genotype (70%) [27]. On the contrary, it is reasonable to speculate that subjects of African and Asian descent would have little risk of suffering panniculitis, for the low incidence of AATD in these populations. Clearly, more epidemiological studies to corroborate or reject these impressions are needed.

Clinically, panniculitis associated with AATD is characterized by nodular lesions and inflammatory subcutaneous, painful, migratory, and recurrent plates, preferably located in proximal regions of the limbs, abdomen, and buttocks, but can also affect the perineum, hips, back, chest, face, and distal parts of the legs (Fig. 12.2). Some lesions regress spontaneously, but others evolve to chronicity and tend to ulcerate and download an oily fluid (which is fat from necrotic adipocytes), leaving permanent scars. Fistulation and discharge of oily liquid are

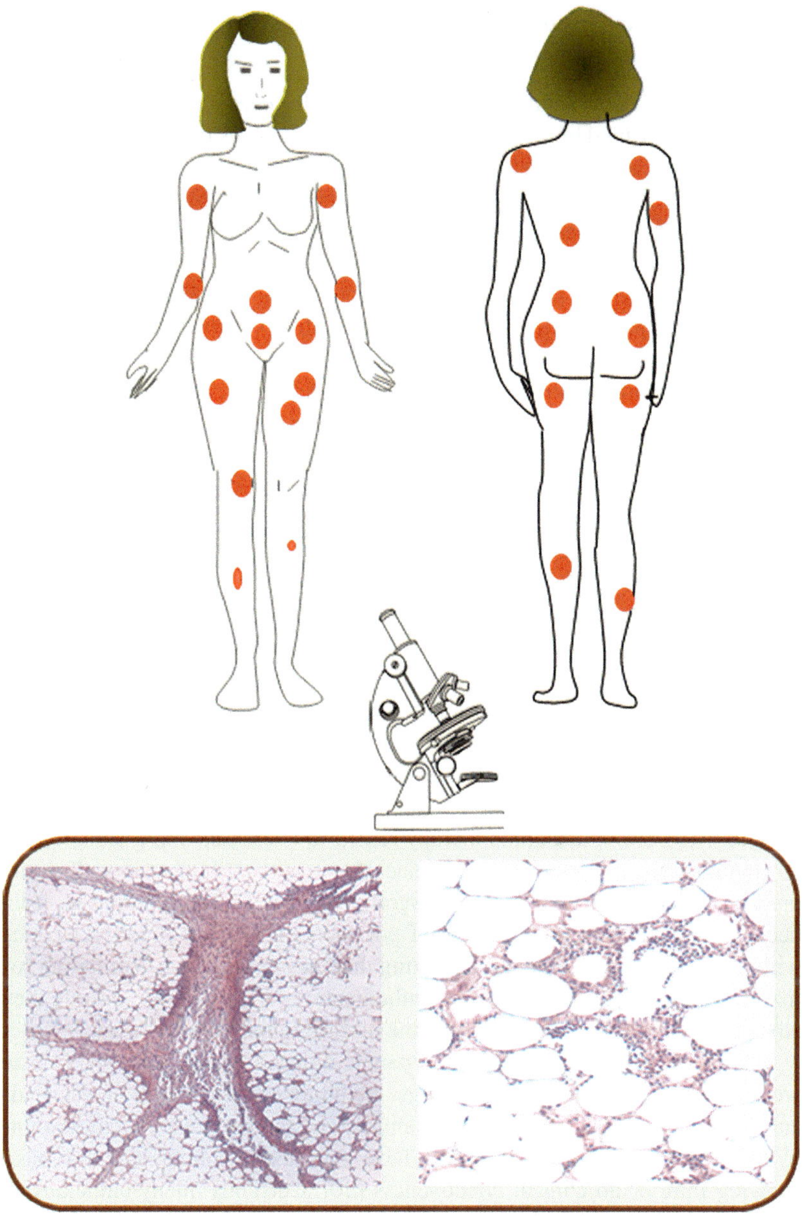

Neutrophils infiltrating the collagen and elastic fibers of the septa surrounding adipocytes, with rupture zones in several septa.

FIGURE 12.2 Anatomical locations of skin lesions and biopsy findings of neutrophilic panniculitis. Neutrophilic panniculitis lesions consist of migratory and recurrent plates, preferably located in proximal regions of the limbs, abdomen, and buttocks, but can also affect other skin regions. Some lesions regress spontaneously, but others tend to ulcerate and download an oily fluid leaving permanent scars. Microscopically it is characterized by dense infiltrates of neutrophils in dermis and connective septa, with secondary lobular inflammation.

fairly typical of AATD, but not pathognomonic. In severe cases, the skin lesions may be accompanied by systemic manifestations, such as fever, fatigue, anorexia, weight loss, and involvement of other organs or body systems. Although in 35% of cases the appearance of skin lesions is preceded by trauma (often low intensity), in most cases no triggering factors are known [24,28–30].

To make a diagnosis of the disease, it is necessary to perform a deep skin biopsy, with an adequate sample of subcutaneous tissue [31]. Histopathologically, AATD-related panniculitis is a lobular panniculitis without vasculitis, characterized by an intense neutrophilic infiltration with collagenolysis and elastolysis (Fig. 12.2). Lesions tend to be focal and well delimited, alternating with large areas of normal panicle. On the periphery of the areas with neutrophilic inflammation there are lymphocytes and macrophages, some of these loaded with fat. Occasonally they have been reported dermal deposits of IgM, C3, and Z polymers [31–33].

AATD-related panniculitis is characterized by low AAT levels, high neutrophil elastase load, and a high neutrophil burden in the inflamed subcutaneous tissues. AATD patients have an increased activation of the TNF-α system, as demonstrated by increased membrane TNF-α and TNFR1 plasma concentrations, and neutrophil-released secondary and tertiary granule antigenic proteins. In this regard, the incidence of autoantibodies directed against degranulated lactoferrin and surface protein accessible to these antibodies is increased in Pi*ZZ patients, leading to an enhanced rate of reactive oxygen species production. In addition, neutrophils are a major source of IL-1β in inflammatory diseases. Interestingly, AAT augmentation therapy results in decreased expression of TNF-α, granule antigenic proteins, and antilactoferrin autoantibodies, as well as suppression of release of IL-1β from several cell types [3,34]. In addition, it has been described the presence of Z-AAT polymers on the skin of a Pi*ZZ patient with neutrophilic panniculitis, suggesting that they could play a facilitating role in the initiation and maintenance of inflammation [33]. However, despite these advances, the precise mechanisms of the disease remain unknown.

The pharmacological treatment of panniculitis associated with AATD is not well established. Several types of drugs, including corticosteroids, nonsteroidal antiinflammatory drugs, cyclophosphamide, colchicine, danazol, chloroquine, hydroxychloroquine, and antibiotics, have been used without scientific basis, showing little or no clinical effectiveness [20]. Currently, many clinicians advocate using dapsone (a diaminodiphenyl sulfone which inhibits the cluster of differentiation molecule CD11B, preventing neutrophil vascular adhesion and migration) as the initial treatment, based on its acceptable effectiveness and reasonable cost [20].

However, augmentation AAT therapy raising AAT serum levels above the 60 mg/dL "protective level" seems the most logical approach because (1) AAT is a potent inhibitor of neutrophil elastase and other proteases present in the inflammatory focus of panniculitis; (2) just like dapsone, inhibits CD11B, preventing neutrophil adhesion and migration; (3) is an excellent inhibitor of IL-1,

IL-1β, TNF-α, and other proinflammatory cytokines; (4) modulates activation, migration, degranulation, and apoptosis of neutrophils, as well as other cells of the immune system involved in inflammation; and (5) AAT infusions have proven highly effective, well tolerated, and without adverse side effects when given off-label as last resort, in a number of patients with AATD–panniculitis who did not respond to conventional therapies [20].

Liver transplantation for advanced cirrhosis normalizes the serum AAT level and has also proved very effective in cases of AATD with panniculitis [35,36], providing further indirect evidence supporting augmentation for episodes of relapsing panniculitis.

12.4 FIBROMYALGIA

Fibromyalgia syndrome (FMS) is a complex chronic pain syndrome affecting approximately 2% of the population (mainly females) worldwide. FMS is characterized by widespread soft-tissue pain, generalized tender points, abnormal fatigue, nonrestorative sleep, and a variety of additional symptoms. Although its precise pathogenesis remains elusive, it involves alterations in multiple ascending and descending central nervous system (CNS) pathways, as well as peripheral pathways, leading to heightened pain sensitivity. No analytical test or imaging techniques for objective diagnosis are currently available [37], and thus an exclusion diagnosis must be used when other typified diseases are ruled out after an appropriate evaluation study in patients fulfilling the American College of Rheumatology criteria for FMS [38]. Another troubling aspect of FMS is the lack of effective therapy for controlling its symptoms, which makes it a major source of personal, family, and social disturbances, and leads to heavy use of health care services, increased work absenteeism, disability, and early retirement [39].

Of note, there are three cases described in the medical literature (two Pi*ZZ and one Pi*MZ middle-aged women) with a rapid and sustained positive response to long-term infusions of AAT, which is still being observed currently [40,41].

In addition, two different studies have demonstrated a higher prevalence of FMS in AATD patients compared with controls, and remarkably FMS symptoms are more severe and appear at earlier ages in AATD than in non-AATD individuals [42,43]. However, a study in Israel in 87 Jewish FMS patients and 200 healthy controls failed to identify significant associations between FMS and AATD, although their results may be distorted by the small number of subjects studied in a population where AATD is extremely rare [44].

To assess the clinical effect of a human plasma-derived AAT concentrate in reducing pain severity of patients with FMS without congenital AATD, 13 subjects with FMS (all with normal serum levels of AAT and normal Pi phenotypes) completed a randomized, double-blind, placebo-controlled, crossover study which did not show a significant improvement over placebo in reducing

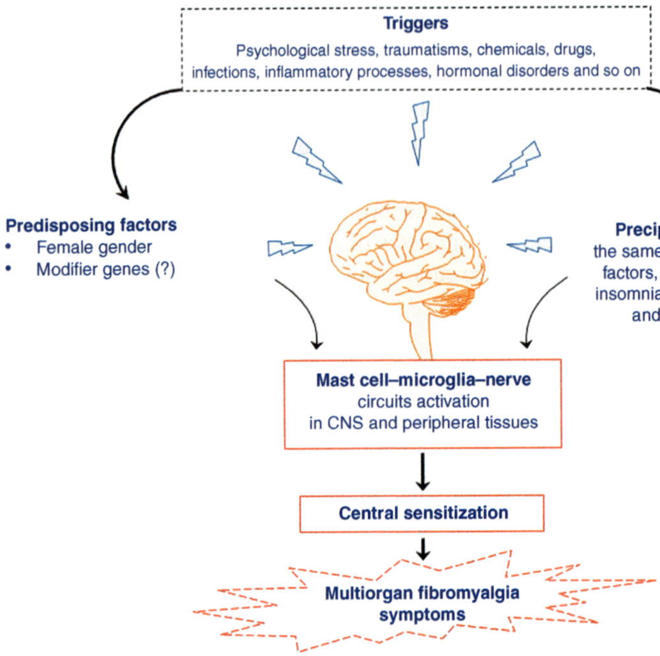

FIGURE 12.3 **Hypothetical mechanisms of fibromyalgia.** According to this scheme, fibromyalgia syndrome would be the consequence of an abnormal activation of mast cells, glia, and nerves neuro–immuno–endocrine circuits in central nervous system and peripheral tissues, by different triggering stimuli, such as infections, traumas, psychological stress, environmental toxics, and so on, and other often unknown, especially in predisposed subjects (e.g., women), with or without the concurrence of genetic modifiers (e.g., antitrypsin deficiency). Activation of mast cell–glia–nerve circuits would cause neuroinflammation and central hypersensitisation, finally causing widespread functional alterations, manifesting by a variety of multisystem symptoms. Subsequently, the same triggers, even at low concentrations, and other normally innocuous stimuli, can act as precipitants of exacerbations and as maintainers of a process with a marked tendency to become chronic.

pain severity and other symptoms of FMS. [45]. Unfortunately, no other similarly designed study in patients with AATD and FMS has been performed.

In the last decade there has been an increasing interest on the likely role of mast cells (MCs) in the pathogenesis of FMS [46,47]. MCs are distributed throughout the human body, including CNS, and have a close relationship with blood vases, glia cells, and nerves [48]. MCs are easily stimulated by multiple factors, including psychological stress and a number of physical, chemical, and infectious triggers, to degranulate releasing various proinflammatory products, such as histamine, proteinases from the granules, eicosanoids from the membrane, cytokines, and so on (Fig. 12.3), able to generate neuroinflammation [49] and secondarily CNS chronic sensitization [50]. Neuroendocrine immune stimuli can induce an increasing number of local MCs and further the extent of their degranulating activity [51]. Microglia cells are activated and respond to proinflamatory signals released by other nonneuronal cells, such as CNS-resident MCs, contributing to spreading the sensation of pain and maintain its chronic status [52].

Given the role of MCs in peripheral and central nociception, and the minimal side effects of ketotifen (an MC stabilizer), a Phase 1 Randomized Controlled Clinical Trial has been performed, and although initial results show no significant changes in the symptoms of patients, researchers are in favor of continuing with increasing doses of the drug [53].

"In vitro," AAT efficiently inhibits histamine, tryptase, and chymase from MCs, in a concentration-dependent manner, suggesting a protective role in MCs-related diseases [54–56]. However, it is not known whether AAT can cross the blood–brain barrier in the FMS related to the AAT to perform their antiinflammatory capabilities in a CNC with possible neuroinflammation.

12.5 RHEUMATOID ARTHRITIS

Rheumatoid arthritis is a chronic autoimmune disorder that primarily affects joints, and is characterized by synovitis, systemic inflammation, and autoantibodies (particularly to rheumatoid factor and citrullinated peptide). Its etiopathogenesis is not fully understood, but involves a combination of genetic and environmental factors. Rheumatoid arthritis affects 0.5%–1.0% of adults, typically women and elderly people. Uncontrolled active rheumatoid arthritis causes joint damage, disability, decreased quality of life, and cardiovascular and other comorbidities. There is no curative treatment for the disease, and available therapy is directed to controlling inflammation and symptoms [57].

With the hypothesis that reduced concentrations of AAT may be a contributory factor to the development and tissue destruction of rheumatoid arthritis, Cox and Huber conducted a pioneering study on the genetic types of AAT in 55 adults with rheumatoid arthritis and 56 children with juvenile rheumatoid arthritis, comparing them with those of 211 healthy adults and 512 schoolchildren matched controls. In this study, Pi types MZ and SZ were significantly more prevalent in adults with rheumatoid arthritis than in controls, but there was no difference between patients with juvenile rheumatoid arthritis and control children [58]. A few years later, these same authors studied 108 patients with rheumatoid arthritis and 98 patients with juvenile rheumatoid arthritis, founding that MZ and SZ heterozygotes were significantly higher among patients than in a control adult population (9.2% vs. 3.5%); however, although among the young there was 4.1% of heterozygotes compared with 1.3% of controls, the difference between the two groups was not statistically significant. In adults, heterozygotes were most frequent among female patients with an early onset of disease. In view of these results, the authors concluded that reduced concentrations of AAT in Z heterozygotes may be inadequate to inhibit the proteolytic enzymes released into the joints of adults with rheumatoid arthritis during phagocytosis of immune complexes [59].

After these two initial Canadian studies, another study was conducted in Sweden in 100 female and 100 male patients with rheumatoid arthritis, finding a significant increase of MZ, MF, and SZ heterozygotes more pronounced in male patients [60].

In contrast to the above, a subsequent study carried out in the United Kingdom in 144 patients with rheumatoid arthritis and 223 normal controls found that the frequencies of F, S, and Z variants were similar in both groups, and no relationships were found between Pi types and rheumatoid disease severity or autoantibody titre [61].

A later British study in 225 adult white patients with common rheumatoid arthritis, 60 patients with severe rheumatoid arthritis, 17 sibling pairs—HLA identical but discordant for rheumatoid arthritis, and 122 random patients with Sjögren's syndrome did not find significant increase in non-M phenotypes in either of the groups of patients with rheumatoid arthritis, but there was a significant increase in M1M2 phenotype and M2 allotype in both the rheumatoid arthritis and severe rheumatoid arthritis groups. No association of variant phenotype was found in four subgroups of patients with Sjögren's syndrome [62].

On the contrary, AAT polymorphisms were studied in 99 subjects with rheumatoid arthritis alone, 21 subjects with rheumatoid arthritis and pulmonary fibrosis, 26 subjects with rheumatoid arthritis and bronchiectasis, and 86 controls, and no associations with either rheumatoid arthritis itself or with the associated pulmonary disorders were noted in this UK Caucasoid population [63].

In addition, another British study in 104 subjects with rheumatoid arthritis without any extraarticular features (29 of whom had severe arthritis and 31 mild disease) and 64 normal controls did not find evidences that Pi*Z either increases susceptibility to rheumatoid arthritis or affects the risk of pulmonary complications or the severity of arthritis in subjects with rheumatoid arthritis [64].

Therefore, the available data cannot confirm or dismiss a potential relationship between AATD and rheumatoid arthritis. More studies are still needed to reach firm conclusions on this topic.

12.6 INFLAMMATORY BOWEL DISEASE

IBD is a group of autoimmune diseases of the colon and small intestine, the most representative of which are Crohn's disease and ulcerative colitis. The causes and mechanisms of IBD are unknown, but presumably it is a complex disease which arises as a result of the interaction of environmental and genetic factors leading to immunological responses and inflammation in the intestine. IBD clinically manifests with severe diarrhea, pain, fatigue, and weight loss. Ulcerative colitis is restricted to the mucosa, while Crohn's disease affects the full thickness of the bowel wall. The diagnosis is usually confirmed by biopsies on colonoscopy. Diagnosis and management of IBD require an integrated evaluation of clinical, endoscopic, radiologic, and pathologic features. There is no cure for IBD and treatment is aimed at relieving inflammation, symptoms, and complications [65,66].

A German study of the late 1990s in 135 patients with IBD (74 of them with Crohn's disease and 61 with ulcerative colitis), compared with 752 healthy unrelated controls, found a very low frequency of AATD phenotypes, with no differences between patients and controls [67].

Conversely, shortly after another study performed in Malmoe in 141 patients with ulcerative colitis, compared with 104 controls, showed a significantly increased frequency of heterozygosity for the Pi*Z variant among patients with ulcerative colitis [68].

However, almost simultaneously a French study of 95 consecutive patients with ulcerative colitis and 63 patients with Crohn's disease found 3 Pi*Z alleles among these patients, matching their prevalence with the average of the normal French population [69].

The Mayo clinic published a retrospective study of 10 patients who had AATD and IBD listed as concurrent diagnoses. Seven patients had ulcerative colitis: four were genotype ZZ and three MZ; four of these seven patients had emphysema, two had asthma, and one had chronic bronchitis. Three of the 10 patients had Crohn's disease, 2 had genotype ZZ and 1 MZ; 2 of the 3 patients were long-term cigarette smokers, and both had emphysema. Nine of the 10 patients with ulcerative colitis required surgery following a failure of medical therapy, a fact suggestive of a worse clinical course of AATD individuals [70].

In a subsequent Canadian study of 300 blood samples from 100 Crohn's disease patients, 100 ulcerative colitis patients, and 100 healthy controls, the Pi*Z allele was found to be significantly associated with Crohn's disease but not with ulcerative colitis [71].

Finally, a retrospective review of the first 651 Pi*ZZ or Pi*Z-null patients from the UK National AATD Registry [26] found that the prevalence of IBD (1.5%) was higher than that predicted in the United Kingdom (0.4%). Ten patients had a confirmed diagnosis of ulcerative colitis, and 1 had Crohn's disease. In two cases there was a family history of IBD.

In short, while some of the available data suggest a probable causal relationship between DDAT and IBD, others indicate otherwise, and so this relationship is currently uncertain.

12.7 MISCELLANEOUS

In addition to the pathologies described in previous sections, there is an extensive list of diseases that have been linked to the AATD, including the following: autoimmune hypothyroidism, aneurysms and arterial dissections, diabetes mellitus, psoriasis, chronic urticaria, mesangiocapilar glomerulonephritis, anterior uveitis, pancreatitis, multiple sclerosis, various types of neoplasms (including those of lung, liver, bladder, gallbladder, pancreas, colon, and lymphomas), as well as other autoimmune and inflammatory diseases. Most of these associations are described in the American Thoracic Society/European Respiratory Society statement for the diagnosis and management of individuals with AATD [72] and in several publications after 2003, easily accessible on the Internet, whose analysis exceeds the limits and objectives designed for this manuscript. In general, most of these associations are not sufficiently proven because many of them are just based on isolated observations or in biased case series with a small sample size or an inadequate sample composition; other lack of appropriate control groups; virtually all lack of randomization of the subjects selection; and also of long-term clinical course follow-up.

Therefore, more studies with sufficient statistical power to support or reject these putative associations are still needed.

REFERENCES

[1] Falk RJ, Gross WL, Guillevin L, Hoffman GS, Jayne DR, Jennette JC, et al. Granulomatosis with polyangiitis (Wegener's): an alternative name for Wegener's granulomatosis. Arthritis Rheum 2011;63:863–4.

[2] Comarmond C, Cacoub P. Granulomatosis with polyangiitis (Wegener): clinical aspects and treatment. Autoimmun Rev 2014;13:1121–5.

[3] Bergin DA, Reeves EP, Hurley K, Wolfe R, Jameel R, Fitzgerald S, et al. The circulating proteinase inhibitor α-1 antitrypsin regulates neutrophil degranulation and autoimmunity. Sci Transl Med 2014;6:217ra1.

[4] Reumaux D, Duthilleul P, Roos D. Pathogenesis of diseases associated with antineutrophil cytoplasm autoantibodies. Hum Immunol 2004;65:1–12.

[5] Korkmaz B, Jaillet J, Jourdan ML, Gauthier A, Gauthier F, Attucci S. Catalytic activity and inhibition of wegener antigen proteinase 3 on the cell surface of human polymorphonuclear neutrophils. J Biol Chem 2009;284:19896–902.

[6] Elzouki AN, Segelmark M, Wieslander J, Eriksson S. Strong link between the alpha 1-antitrypsin PiZ allele and Wegener's granulomatosis. J Intern Med 1994;236:543–8.

[7] Lhotta K, Vogel W, Meisl T, Buxbaum M, Neyer U, Sandholzer C, et al. Alpha 1-antitrypsin phenotypes in patients with anti-neutrophil cytoplasmic antibody-positive vasculitis. Clin Sci (Lond) 1994;87:693–5.

[8] Savige JA, Chang L, Cook L, Burdon J, Daskalakis M, Doery J. Alpha 1-antitrypsin deficiency and anti-proteinase 3 antibodies in anti-neutrophil cytoplasmic antibody (ANCA)-associated systemic vasculitis. Clin Exp Immunol 1995;100:194–7.

[9] Baslund B, Szpirt W, Eriksson S, Elzouki AN, Wiik A, Wieslander J, et al. Complexes between proteinase 3, alpha 1-antitrypsin and proteinase 3 anti-neutrophil cytoplasm autoantibodies: a comparison between alpha 1-antitrypsin PiZ allele carriers and noncarriers with Wegener's granulomatosis. Eur J Clin Invest 1996;26:786–92.

[10] Griffith ME, Lovegrove JU, Gaskin G, Whitehouse DB, Pusey CD. C-antineutrophil cytoplasmic antibody positivity in vasculitis patients is associated with the Z allele of alpha-1-antitrypsin, and P-antineutrophil cytoplasmic antibody positivity with the S allele. Nephrol Dial Transplant 1996;11:438–43.

[11] Callea F, Gregorini G, Sinico A, Gonzales G, Bossolasco M, Salvidio G, et al. alpha 1-Antitrypsin (AAT) deficiency and ANCA-positive systemic vasculitis: genetic and clinical implications. Eur J Clin Invest 1997;27:696–702.

[12] Esnault VL, Audrain MA, Sesboue R. Alpha-1-antitrypsin phenotyping in ANCA-associated diseases: one of several arguments for protease/antiprotease imbalance in systemic vasculitis. Exp Clin Immunogenet 1997;14:206–13.

[13] Borgmann S, Endisch G, Urban S, Sitter T, Fricke H. A linkage disequilibrium between genes at the serine protease inhibitor gene cluster on chromosome 14q32.1 is associated with Wegener's granulomatosis. Clin Immunol 2001;98:244–8.

[14] Chorostowska-Wynimko J, Gawryluk D, Struniawski R, Popławska B, Fijołek J. Incidence of alpha-1 antitrypsin Z and S alleles in patients with granulomatosis with polyangiitis—pilot study. Pneumonol Alergol Pol 2013;81:319–22.

[15] Pervakova MY, Emanuel VL, Titova ON, Lapin SV, Mazurov VI, Belyaeva IB, et al. The diagnostic value of alpha-1-antitrypsin phenotype in patients with granulomatosis with polyangiitis. Int J Rheumatol 2016;2016:7831410.

[16] Mahr AD, Edberg JC, Stone JH, Hoffman GS, St Clair EW, Specks U, et al. Alpha$_1$-antitrypsin deficiency-related alleles Z and S and the risk of Wegener's granulomatosis. Arthritis Rheum 2010;62:3760–7.

[17] Morris H, Morgan MD, Wood AM, Smith SW, Ekeowa UI, Herrmann K, et al. ANCA-associated vasculitis is linked to carriage of the Z allele of {alpha}1 antitrypsin and its polymers. Ann Rheum Dis 2011;70:1851–6.

[18] Dowd SK, Rodgers GC, Callen JP. Effective treatment with alpha 1-protease inhibitor of chronic cutaneous vasculitis associated with alpha 1-antitrypsin deficiency. J Am Acad Dermatol 1995;33(5 pt 2):913–6.

[19] Hernández Pérez JM, Fumero García S, Alvarez Pío A. Successful α1-antitrypsin replacement therapy in a patient with α1-antitrypsin deficiency and granulomatosis with polyangiitis. Rheumatology (Oxford) 2013;52:755 7.

[20] Blanco I, Lipsker D, Lara B, Janciauskiene S. Neutrophilic panniculitis associated with alpha-1-antitrypsin deficiency: an update. Br J Dermatol 2016;174:753–62.

[21] Warter J, Storck D, Grosshans E, Metais P, Kuntz JL, Klumpp T. Syndrome de Weber-Christian associé à un déficit en alpha-I-antitrypsine. Enquéte familiale. Ann Med Interne (Paris) 1972;123:877–82.

[22] Grosshans E, Warter J, Storck D. A propos d'une panniculite liquéfiante: Les manifestations cutanées del déficits en alpha-I-antitrypsine. Bull Soc Fr Dermatol Syphiligr 1973;80:244–7.

[23] Rubinstein HM, Jaffer AM, et al. Alpha1-antitrypsin deficiency with severe panniculitis. Report of two cases. Ann Intern Med 1977;86:742–4.

[24] Smith KC, Pittelkow MR, Su WP. Panniculitis associated with severe alpha 1-antitrypsin deficiency. Treatment and review of the literature. Arch Dermatol 1987;123:1655–61.

[25] Stoller JK, Piliang M. Panniculitis in alpha-1antitrypsin deficiency: a review. Clin Pulm Med 2008;15:113–7.

[26] Stone H, Pye A, Stockley RA. Disease associations in alpha-1-antitrypsin deficiency. Respir Med 2014;108:338–43.

[27] Valverde R, Rosales B, Ortiz-de Frutos FJ, Rodríguez-Peralto JL, Ortiz-Romero PL. Alpha-1-antitrypsin deficiency panniculitis. Dermatol Clin 2008;26:447–51.

[28] Cathomas M, Schüller A, Candinas D, Inglin R. Severe postoperative wound healing disturbance in a patient with alpha-1-antitrypsin deficiency: the impact of augmentation therapy. Int Wound J 2015;12:601–4.

[29] Linares-Barrios M, Conejo-Mir IS, Artola Igarza JL, Navarrete M. Panniculitis due to alpha 1-antitrypsin deficiency induced by cryosurgery. Br J Dermatol 1998;138:552–3.

[30] Rajagopal R, Malik AK, Murthy PS, Neog LS. Alpha-1 antitrypsin deficiency panniculitis. Indian J Dermatol Venereol Leprol 2002;68:362–4.

[31] Chan MP. Neutrophilic panniculitis: algorithmic approach to a heterogeneous group of disorders. Arch Pathol Lab Med 2014;138:1337–43.

[32] Geller JD, Su WPD. A subtle clue to the histopathologic diagnosis of early alpha 1-antitrypsin deficiency panniculitis. J Am Acad Dermatol 1994;31:241–5.

[33] Gross B, Grebe M, Wencker M, et al. New findings in PiZZ alpha(1)-antitrypsin deficiency-related panniculitis. Demonstration of skin polymers and high dosing requirements of intravenous augmentation therapy. Dermatology 2009;218:370–5.

[34] Dinarello CA. Immunological and inflammatory functions of the interleukin-1 family. Annu Rev Immunol 2009;27:519–50.

[35] O'Riordan K, Blei A, Rao MS, Abecassis M. alpha 1-antitrypsin deficiency-associated panniculitis: resolution with intravenous alpha 1-antitrypsin administration and liver transplantation. Transplantation 1997;63:480–2.

[36] Fernández-Torres R, García-Silva J, Robles O, et al. Alfa-1-antitrypsin deficiency panniculitis acquired after liver transplant and successfully treated with retransplant. J Am Acad Dermatol 2009;60:715–6.

[37] Smith HS, Harris R, Clauw D. Fibromyalgia: an afferent processing disorder leading to a complex pain generalized syndrome. Pain Physician 2011;14:E217–45.

[38] Chinn S, Caldwell W, Gritsenko K. Fibromyalgia pathogenesis and treatment options update. Curr Pain Headache Rep 2016;20:25.

[39] Lacasse A, Bourgault P, Choinière M. Fibromyalgia-related costs and loss of productivity: a substantial societal burden. BMC Musculoskelet Disord 2016;17:168.

[40] Blanco I, Canto H, de Serres FJ, Fernandez-Bustillo E, Rodríguez MC. Alpha1-antitrypsin replacement therapy controls fibromyalgia symptoms in 2 patients with PI ZZ alpha1-antitrypsin deficiency. J Rheumatol 2004;31:2082–5.

[41] Blanco I, Astudillo A, Domínguez F, Janciauskiene S, Cárcaba V, Gallo C, et al. Intravenous infusions of purified alpha 1-antitrypsin effectively controls symptoms and reverts muscle biopsy changes in an MZ-alpha-1 antitripsyn deficiency and fibromyalgia syndrome patient. J Musculoskelet Pain 2010;18:167–72.

[42] Blanco I, Arbesú D, Al Kassam D, De Serres F, Fernández-Bustillo E, Rodríguez Menéndez C. Alpha1-antitrypsin polymorphism in fibromyalgia patients from the Asturias Province in Northern Spain: a significantly higher prevalence of the PI*Z deficiency allele in patients than in the general population. J Musculoskelet Pain 2006;14:5–12.

[43] Schmechel DE, Edwards CL. Fibromyalgia, mood disorders, and intense creative energy: A1AT polymorphisms are not always silent. Neurotoxicology 2012;33:1454–72.

[44] Ablin JN, Bar-Shira A, Yaron M, Orr-Urtreger A. Candidate-gene approach in fibromyalgia syndrome: association analysis of the genes encoding substance P receptor, dopamine transporter and alpha1-antitrypsin. Clin Exp Rheumatol 2009;27:S33–8.

[45] Alegre C, Barceló M, Jardí R, Rodríguez-Frias F, Camprubí S. α1-Antitrypsin in fibromyalgia: results of a randomized, placebo-controlled, double-blind and crossover pilot trial. Musculoskeletal Care 2012;10:178–83.

[46] Lucas HJ, Brauch CM, Settas L, Theoharides TC. Fibromyalgia—new concepts of pathogenesis and treatment. Int J Immunopathol Pharmacol 2006;19:5–10.

[47] Tsilioni I, Russell IJ, Stewart JM, Gleason RM, Theoharides TC. Neuropeptides CRH, SP, HK-1, and inflammatory cytokines IL-6 and TNF are increased in serum of patients with fibromyalgia syndrome, implicating mast cells. J Pharmacol Exp Ther 2016;356:664–72.

[48] Salamon P, Shoham NG, Gavrieli R, Wolach B, Mekori YA. Human mast cells release interleukin-8 and induce neutrophil chemotaxis on contact with activated T cells. Allergy 2005;60:1316–9.

[49] Kosek E, Altawil R, Kadetoff D, Finn A, Westman M, Le Maître E, et al. Evidence of different mediators of central inflammation in dysfunctional and inflammatory pain—interleukin-8 in fibromyalgia and interleukin-1 β in rheumatoid arthritis. J Neuroimmunol 2015;280:49–55.

[50] Cassisi G, Ceccherelli F, Atzeni F, Sarzi-Puttini P. Reply to: mast cells in fibromyalgia. Clin Exp Rheumatol 2015;33:S140.

[51] Mazzari S, Canella R, Petrelli L, Marcolongo G, Leon A. N-(2-hydroxyethyl)hexadecanamide is orally active in reducing edema formation and inflammatory hyperalgesia by down-modulating mast cell activation. Eur J Pharmacol 1996;300:227–36.

[52] Hansson E. Could chronic pain and spread of pain sensation be induced and maintained by glial activation? Acta Physiol (Oxf) 2006;187:321–7.

[53] Ang DC, Hilligoss J, Stump T. Mast cell stabilizer (ketotifen) in fibromyalgia: phase 1 randomized controlled clinical trial. Clin J Pain 2014; [Epub ahead of print].

[54] He SH, Xie H, Zhang XJ, Wang XJ. Inhibition of histamine release from human mast cells by natural chymase inhibitors. Acta Pharmacol Sin 2004;25:822–6.

[55] He SH, Xie H. Modulation of histamine release from human colon mast cells by protease inhibitors. World J Gastroenterol 2004;10:337–41.

[56] He SH, Chen P, Chen HQ. Modulation of enzymatic activity of human mast cell tryptase and chymase by protease inhibitors. Acta Pharmacol Sin 2003;24:923–9.

[57] Scott DL, Wolfe F, Huizinga TW. Rheumatoid arthritis. Lancet 2010;376:1094–108.

[58] Cox DW, Huber O. Rheumatoid arthritis and alpha-1-antitrypsin. Lancet 1976;1:1216–7.

[59] Cox DW, Huber O. Association of severe rheumatoid arthritis with heterozygosity for alpha 1-antitrypsin deficiency. Clin Genet 1980;17:153–60.

[60] Beckman G, Beckman L, Bjelle A, Dahlqvist SR. Alpha-1-antitrypsin types and rheumatoid arthritis. Clin Genet 1984;25:496–9.

[61] Sanders PA, Thomson W, Browne DA, Dyer PA, Grennan DM. Alpha 1 antitrypsin (PI) allotypes in rheumatoid arthritis. Ann Rheum Dis 1986;45:450–3.

[62] Papiha SS, Pal B, Walker D, Mangion P, Hossain MA. Alpha 1 antitrypsin (PI) phenotypes in two rheumatic diseases: a reappraisal of the association of PI subtypes in rheumatoid arthritis. Ann Rheum Dis 1989;48:48–52.

[63] Bate AS, Sidebottom D, Cooper RG, Loftus M, Chattopadhyay C, Grennan DM. DNA variants of alpha-1-antitrypsin in rheumatoid arthritis with and without pulmonary complications. Dis Markers 1990;8:317–21.

[64] Steers G, McMahon MJ, Grennan DM, Hillarby MC. Lack of association of the alpha-1-antitrypsin PIZ allele with rheumatoid arthritis or with its extra articular complications. Dis Markers 1992;10:151–7.

[65] Ordás I, Eckmann L, Talamini M, Baumgart DC, Sandborn WJ. Ulcerative colitis. Lancet 2012;380:1606–19.

[66] Baumgart DC, Sandborn WJ. Crohn's disease. Lancet 2012;380:1590–605.

[67] Folwaczny C, Urban S, Schröder M, Hofmann B, Noehl N, König A, et al. Alpha1-antitrypsin alleles and phenotypes in patients with inflammatory bowel disease. Scand J Gastroenterol 1998;33:78–81.

[68] Elzouki AN, Eriksson S, Lofberg R, Nassberger L, Weislander J, Lindgren S. The prevalence and clinical significance of alpha1-antitrypsin deficiency (PiZ) and ANCA specificities (proteinase 3, BPI) in patients with ulcerative colitis. Inflamm Bowel Dis 1999;5:246e52.

[69] Taddei C, Audrain MA, Reumaux D, Sesboüe R, Testa A, Galmiche JP, et al. Alpha1-antitrypsin phenotypes and anti-neutrophil cytoplasmic auto-antibodies in inflammatory bowel disease. Eur J Gastroenterol Hepatol 1999;11:1293–8.

[70] Yang P, Tremaine WJ, Meyer RL, Prakash UB. Alpha 1-antitrypsin deficiency and inflammatory bowel diseases. Mayo Clin Proc 2000;75:450e5.

[71] Kotlowski R, Bernstein CN, Silverberg MS, Krause DO. Population-based case-control study of alpha 1-antitrypsin and SLC11A1 in Crohn's disease and ulcerative colitis. Inflamm Bowel Dis 2008;14:1112–7.

[72] American Thoracic Society European Respiratory Society. American Thoracic Society/European Respiratory Society statement: standards for the diagnosis and management of individuals with alpha-1 antitrypsin deficiency. Am J Respir Crit Care Med 2003;168:818–900.

Chapter 13

Clinical Management and Treatment of Lung Disease

13.1 INITIAL EVALUATION OF THE PATIENT

In general, the assessment of a patient with COPD should always include assessment of symptoms, severity of airflow limitation, history of exacerbations, and comorbidities. In clinical practice, the physician who assesses a patient with confirmed AATD for the first time should get a detailed clinical history, including the family and personal history, specifically asking by smoking and occupational exposure to toxic fumes, suspicious symptoms of AATD in other family members, onset and progression of symptoms, previous diagnoses and treatments, current status, and quality of life. In addition, the physical examination will be complete, looking for signs of COPD, liver involvement, and any other abnormal finding [1].

The baseline assessment should include a lung function with spirometry and lung volumes (before and after an inhaled bronchodilator), a diffusing capacity of the lungs for carbon monoxide (DLCO) and a diffusing capacity per unit volume standardized for alveolar volume (KCO). It is essential to measure the oxygenation of the patient through a pulse oximetry and an arterial gas analysis if oxygen saturation is approximately 90% or lower. A chest radiograph should be performed because although it may be little sensitive to assess a nonsevere COPD, it is always useful to rule out other diseases. If the simple radiograph of the chest is normal or unspecific, but the patient refers symptoms suggestive of other non-detectable lung diseases with simple chest x-ray (e.g., bronchiectasis, or hidden lung cancer) performing a computed tomography (CT) should be considered, but repeating this exploration, if no clear indication, it should be avoided whenever possible, due to the high exposure to ionizing radiation that this exploration entails [2].

A comprehensive clinical panel of blood and urine chemistry and haematology tests (including always hepatic enzymes to rule out liver disease), should always be ordered to determine the patient's general health status. Once the diagnosis of AATD-COPD is confirmed, it is convenient to classify the disease severity using international agreed criteria, such as those established by the Global Initiative for Chronic Obstructive Lung Disease, to apply the recommended treatment with the amendments considered appropriate for each individual case [3,4].

Asymptomatic nonsmokers or former smokers AATD subjects with normal respiratory function do not require any pharmacological treatment, but they should avoid exposure to tobacco smoke and other environmental contaminants. In the case that the subject is yet an active smoker, the physician should inform him about the high risks associated with smoking, and seriously advise him to quit smoking while offering help with psychological and pharmacological treatments to achieve this important goal.

If the physician who makes the diagnosis has concerns about management, referring the patient to other specialist experienced in the management of patients with AATD should be seriously considered [5].

13.2 GENETIC COUNSELING

Siblings of individuals with AATD are the only family members for whom genetic testing is strongly recommended because they have 25% or even more risk of also having this hereditary condition [6].

The parents of an individual ZZ are usually heterozygotes MZ or SZ, and less frequently one of them may be homozygous for the Z allele. When both parents are MZ heterozygotes, each sibling of an affected individual has a 25% chance of being ZZ, a 50% chance of being MZ, and a 25% chance of being MM. In the less common circumstance that one parent is ZZ homozygote and the other MZ heterozygote, the risk to each sibling of being ZZ is 50%, and the chance of being MZ is also 50% [5].

Although the risk of AATD in parents, offspring, and second-degree relatives of patients is low but greater than that of the general population, it is a common practice to discuss with them the benefits and risks of AAT testing prior to its realization [7]. Carrier testing for at-risk family members and prenatal testing for pregnancies at increased risk are possible once the pathogenic SERPINA1 variants have been identified in the family [5].

13.3 STANDARD THERAPY FOR COPD

AATD individuals diagnosed with COPD should be treated with the common pharmacological and nonpharmacological measures used for any patient suffering from COPD [1–5]. Vaccination against influenza and pneumococcal pneumonia is recommended. Treatment of the disease during periods of clinical stability must be comprehensive and progressive depending on symptoms, degree of airways obstruction, and frequency of exacerbations [3]. During exacerbations there is an excess of proteases and a deficient antiprotease screen within lung alveoli [8], so an early and aggressive treatment is recommended, with increasing doses of bronchodilators, short courses of corticosteroids and antibiotics.

Respiratory rehabilitation, if available and indicated, should be offered to selected patients with functional impairment. Long-term oxygen therapy will be considered for patients who meet the established criteria to receive it.

Lung transplantation is a treatment option for selected patients with end-stage lung disease [9]. According to the 2007 report of the International Society for Heart and Lung Transplantation, 8% of lung transplants were performed in patients with a severe AATD [10]. As for lung volume reduction surgery in patients with AATD, experience is limited and the available results of this surgery inconclusive [11]. However, in selected cases it could be considered as a bridge to lung transplantation [12].

13.4 AUGMENTATION THERAPY WITH INTRAVENOUS INFUSIONS OF HUMAN PLASMA-DERIVED ALPHA-1 ANTITRYPSIN

13.4.1 Historical Perspective

13.4.1.1 Preliminary Advances

The first practical large-scale method of blood plasma fractionation was developed by Edwin J. Cohn (1892–1953) in the United States at the beginning of the 1940s. The Cohn process consisted in a series of ethanol precipitation techniques with the purpose of extracting albumin from blood plasma, in principle directed to treat wounded soldiers in World War II. In the following years, the Cohn process served to fractionate and remove from the human plasma several essential proteins other than albumin, such as gamma globulin, fibrinogen, clotting factors, or AAT [13,14].

The AAT augmentation therapy history begins in 1979 when James Gadek joined the Ronald Crystal group of the Pulmonary Branch at the National Institutes of Health (NIH) in Bethesda. Gadek had recent experience in augmentative therapy in the field of hereditary C1 esterase deficiency, and knew that the administration of purified C1 inhibitor was useful to treat acute episodes of angioedema [15]. At that time the Crystal team was investigating the AAT behavior in the alveolar fluid obtained by bronchoalveolar lavage from AATD-COPD patients, observing that they had abnormally reduced levels of the protein in the alveolar fluid [16]. With these observation, the Crystal's research group hypothesized that the emphysema associated with inherited serum deficiency of AAT appeared to result from an imbalance between neutrophil elastase and AAT within pulmonary alveoli, and assessed the feasibility of reversing this biochemical defect with parenteral AAT concentrates of normal plasma generated by sequential precipitations with polyethylene glycol. Then, they treated five Pi*Z patients with advanced emphysema, with four intravenous weekly infusions of 4 g of AAT. During this period AAT serum levels were maintained more than twice the basal, and bronchoalveolar lavage demonstrated an effective antielastase activity within the alveolar structures. Of note, there were no adverse side effects, and no subjects developed antibody responses to the infused AAT [17].

Next, the Crystal team continued research with 21 severely affected AATD patients who were treated with 60 mg of plasma-derived AAT per kilogram of body weight, once a week for up to 6 months, demonstrating that infusions of AAT reversed the biochemical abnormalities in serum and lung fluid of treated patients. No changes in lung function were observed, and the only important adverse reactions were four episodes of self-limited fever. This study served to confirm the suitability of the 60 mg/kg/week dose, which is still used since that time [18].

The results of this pioneering study were soon replicated by subsequent studies. Thus, in a 6-month multicenter study conducted in the Federal Republic of Germany, with 20 COPD-Pi*Z patients treated with an intravenous weekly dose of 60 mg/kg of human plasma-derived AAT, it was found that this dose raised AAT serum levels above a protective level of 35%, and also corrected the deficient levels of AAT in the bronchoalveolar fluid. Of note, the therapy was well tolerated, without significant side effects and without developing antibodies against the protein [19].

In parallel, in an attempt of reducing the infusions frequency the Crystal group evaluated the ability of 250 mg/kg of AAT infusions at 28-day intervals for 12 months to 9 individuals with AATD and emphysema, finding that serum AAT levels exceeded the protective threshold for an average of 25 days after each dose. Furthermore, the postinfusion level of AAT in the nadir lung epithelial lining fluid was fivefold greater than the preinfusion level, and the anti-neutrophil elastase capacity of the nadir epithelial lining fluid was also elevated approximately threefold above the preinfusion level [20].

13.4.1.2 Creation of the American National Registry of Alpha-1 Antitrypsin Deficiency

Following the Crystal et al. preliminary clinical studies, and with the development of large-scale purified human AAT (later known under the trade name of "Prolastin"), the US Food and Drug Administration (FDA) approved the use of AAT augmentation therapy for adults with severe AATD-associated pulmonary emphysema in 1987, and this therapy began to spread gradually to many other countries around the world. The approval was supported by the Orphan Drug Act of 1983 (a law passed in the United States to facilitate development of orphan drugs for rare diseases), and the assumption that a long-term randomized placebo-controlled trial was practically impossible to carry out by the shortage of individuals with the disease, the likely long duration and the foreseeable complexity that would be the analysis of the results [16].

As a condition for approving the license of Prolastin, the FDA demanded the creation of a registry of patients, treated and untreated, to define the natural history and clinical course of this disorder, as well as to monitor the therapy safety in the long term. Therefore, in 1988, the NIH National Heart, Lung, and Blood Institute (NHLBI) initiated a registry of individuals with AATD, which involved 36 clinical centers in the United States and 1 center in Canada, between March 1989 and October 1992. The registered population was constituted by a total of 1129 AATD adult subjects, having serum AAT levels ≤11 μM/L. Of these recruited subjects 309 were always on Prolastin, 357 were discontinuously treated with Prolastin, and 382 were not treated with augmentation therapy [21,22]. The comparative analysis of the groups did not find overall difference in the rate of FEV_1 loss, although it was observed a significant lower loss in treated vs. untreated subjects who had a FEV_1 between 35% and 49%. In addition, the relative risk of mortality was significantly lower in the treated group. Therefore, although the study was not randomized, their results supported the idea that augmentation therapy was safe and beneficial to raise levels of AAT in blood and lung interstitium to prevent the proteolytic destruction of the lungs by neutrophil elastase and to slow down the emphysema progression.

13.4.2 The Theoretical Protective Threshold Value

Normal AAT serum levels are between 20 and 53 micromolar units per liter (abbreviated as μmol/L or μM/L), equivalent to approximately 90–200 mg/dL measured by nephelometry.

On the contrary, serum levels observed in AATD with lung disease are usually <11 μM/L (equivalent to approximately <60 mg/dL measured by nephelometry, or what is the same 80 mg/dL determined by radioimmunodiffusion). These serum levels were empirically considered as the cut-off value at which AAT protects lung tissues against neutrophil elastase, and therefore below which the risk of developing emphysema increases. This threshold value, although empirical, has its basis in clinical practice because individuals with values below this threshold are at increased risk of developing emphysema, while phenotypes whose levels of AAT exceed this level are usually free from emphysema [23–27].

This proposed value is important because in clinical practice it can be used to decide, in a simplistic way, whether prescribe augmentative therapy to a patient. Besides, there is some confusion about the equivalence between the values obtained with the pure standard developed by the US NIH, which are expressed in micromolar units (i.e., μmol/L or μM), and those using the nonpurified standard, which are expressed as milligrams per deciliter (mg/dL). In 2003, the ATS/ERS statement set up that a "protective" threshold level of 11 μmol/L corresponds to 80 mg/dL if measured by radial immunodiffusion and to 50 mg/dl if measured by nephelometry [1,4].

Radial immunodiffusion overestimates the AAT concentration by as much as 35%–40%, and although less nephelometry may also overestimate (although less) AAT levels because of interference with lipids or hemoglobin [1]. Thus, in later years, radial immunodiffusion is being replaced in many laboratories by the nephelometry method, and the proposed protective threshold value of 11 micromolar is being translated as equivalent to 57 mg/dL (measured by nephelometry) [1,5,28], and even a conversion factor of 5.2 to convert micromolar to milligrams per deciliter has been provided (i.e., $11 \times 5.2 = 57.2$) [29]. On the contrary, to convert 11 μmol/L from the International System of Units

(SI) to conventional units (mg/dL), the value in micromoles is divided by the conversion factor 0.184; that is, $(11/0.184) = 59.8\,\text{mg}/\text{dL}$ (http://www.globalrph.com/conv_si.htm). Therefore, it seems that for orientation purposes the 11 μmol/L equivalence for mg/dL may be between 50 and 60, but possibly closer to the latter.

13.4.3 Clinical Efficacy of the Augmentation Therapy

Since the introduction of augmentation treatment for clinical use, findings from observational and cohort studies have shown that the rate of FEV_1 loss is slower in individuals who receive augmentation treatment than in those who do not, and besides this therapy was associated with reduced mortality in patients with moderate-to-severe AATD-COPD [30–34].

However, as in contrast to the well-planned, randomized clinical trials, these observational studies are subject to a number of potential disadvantages that may bias their results (e.g., selection bias age, socioeconomic status, information bias, measurement error, confounders, and further factors) [35], and their findings have not been considered sufficiently convincing when analyzed with statistical criteria of evidence-based medicine. Besides, as the natural evolution of emphysema is too slow to observe significant changes in FEV_1 in the short or medium term, it has been estimated that a cohort of more than hundred people would be needed to be randomized to treatment or placebo for 5 years to establish the effect of treatment [36]. Therefore, this endpoint has been considered unhelpful because the rarity of the disease and the associated difficulties in finding patients, the high cost of developing a clinical trial with this methodology, and the ethical problems involved. Therefore, these difficulties led researchers to explore more sensitive alternative methods than FEV_1 that would make possible a definitive randomized, placebo-controlled trial in fewer patients for less time, choosing as outcome to measure the lung density quantified by CT, which better shows lung destruction than do the traditional measurements of lung function.

Thus, Dirksen et al. in 1999 analyzed FEV_1, DLCO, and CT lung density endpoints in a randomized, placebo-controlled trial conducted in Denmark, reporting slower rates of lung density loss in patients given augmentation therapy than in those given placebo (2.6 g/L/year with placebo and 1.5 g/L/year with therapy), although the difference was not significant ($P = .07$) by the small sample size (56 patients) and insufficient duration of the trial [37].

Ten years later, the same author conducted a new trial, designated as EXACTLE (EXAcerbations and Computed Tomography scan as Lung End-points), in which the CT densitometry was used as the primary outcome measure, in a total of 77 Pi*Z subjects randomized to weekly infusions of 60 mg/kg of AAT or placebo for 2–2.5 years. In this trial, CT showed a suggestive trend of treatment benefit (*P* values ranged .049–.084, and a reduction in exacerbation severity was observed) [38].

More recently, in July 2015 the results of the first well-powered multicenter, double-blind, randomized, placebo-controlled trial of AAT treatment in patients with severe AATD-COPD, designated as RAPID (Randomized, Placebo-controlled Trial of Augmentation Therapy in Alpha-1 Proteinase Inhibitor Deficiency), were published [39] (Table 13.1). Briefly, the conclusions of the trial can be summarized as follows: (1) AAT therapy significantly reduced the annual lung density rate of decline measured by CT scan at total lung capacity in treated subjects vs. placebo (0.74 g/L/year, or 34%). (2) Patients switched from placebo to AAT therapy experienced a remarkable reduction in the progression of emphysema. (3) The late introduction of treatment reduced the decline of lung density, but delaying treatment resulted in irreversible loss of lung tissue, a fact to consider for starting an earlier treatment. (4) A dose–response relationship (i.e., the more plasma concentration of AAT, the less loss of lung density) was observed, and this recommends further studies aimed at determining the optimal dose of AAT to achieve the desired protective effect. (5) There were no differences in adverse effects between the treated and the placebo groups. (6) As expected, it was not possible to analyze other secondary objectives (i.e., exacerbations, lung function, or quality of life) by the excessive withdrawal of patients in the placebo group, explicable by the ineffectiveness of placebo (Fig. 13.1).

13.4.4 Clinical Indications for Intravenous Augmentation Therapy

Intravenous augmentation therapy is the only specific therapy for patients with severe congenital AATD, and should be assessed by medical experts for AATD individuals with established fixed airflow obstruction [1]. This therapy is currently available in USA, Canada, Germany, Austria, Poland, Spain, Italy, Ireland, France, Belgium, Switzerland, the Netherlands, Denmark, Sweden, Norway, Finland, Greece, Portugal, Argentina, Brazil, Chile, Colombia, and Mexico [40].

In general, all currently available guidelines recommend augmentation therapy for individuals with abnormal SERPINA1 genotypes expressing AAT serum levels below 11 μM (equivalent to approximately 50–60 mg/dL by nephelometry) and documented evidence of airflow obstruction in pulmonary function tests [1,41–43]. Its aim is to raise baseline serum AAT concentrations above the protective levels, to delay the progression of pulmonary emphysema, to slow down the lung function deterioration, to reduce rate and severity of exacerbations, and to improve the patient's quality of life.

Initial recommendations of the 1989-ATS guide [41] and the 2003-ATS/ERS statement [1], based on the available scientific evidence, recommend treatment for individuals with serum concentrations below 11 μM and established fixed airflow obstruction (FEV_1 less than 80%), especially when FEV_1 post-bronchodilation is from 35% to 60% of predicted. In addition, subjects with normal or nearly normal pulmonary function can be treated if they experience a rapid decline in lung function (e.g., $\Delta FEV_1 > 120$ mL/year).

TABLE 13.1 Summary of Selected Trials on Augmentation Therapy

Author, year [reference]	Type of study	Results
No randomized studies		
Seersholm, 1997 [30]	Observational. *N* = 295 Pi*Z COPD: 198 treated (60 mg/kg/week) vs. 97 untreated. Average follow-up time of 3.2 years in the treated group and 5.8 years in the untreated	Reduction in FEV1decline in the treated group, with greater benefit in patients with FEV_1 31%–65%
The Alpha-l-Antitrypsin Deficiency Registry Study Group, 1998 [31]	Observational with control group (*n* = 1129) 33% of cases 60 mg/kg/week; 43%, 120 mg/kg/14 days; and 24%, 180 mg/kg/month	Significant decrease in mortality; significant smaller FEV_1decline among patients in augmentation therapy and FEV_1 35%–49%
Wencker, 2001 [32]	Observational without control group (n = 96). FEV_1 decline (pre–post); 60 mg/kg/week	Significant smaller decline of FEV_1 during the therapy period. Smaller decline in patients with FEV_1 >65%
Tonelli, 2009 [33]	Observational; 164 patients enrolled in the Alpha-1 Foundation DNA and Tissue Bank study, treated (60 mg/kg/week) vs. 40 untreated. Mean follow-up time 41.7 months	Smaller decline of FEV_1 in ex-smokers with FEV_1 <50%
Randomized placebo-controlled studies		
Dirksen, 1999 [37]	56 COPD patients with FEV_1 30%–80%, randomized to either AAT (250 mg/kg/month) or albumin, for at least 3 years	Loss of pulmonary tissue measured by CT of 2.6 g/L/year with placebo and 1.5 g/L/year with therapy (*P* = .07). No differences in FEV_1 and DLCO decline
Dirksen, 2009 [38]	EXACTLE study: 77 COPD with FEV_1 25%–80%, 35 treated 60 mg/kg/week vs. 32 placebos, 30 months. Pulmonary density measured by CT	Almost significant reduction in lung loss density in treated vs. placebo (*P* = .049). No differences in FEV_1 and DLCO decline, neither in number of exacerbations, but they were milder among treated
Chapman, 2015 [39]	RAPID study: 153 COPD with FEV_1 35%–70%: 84 treated (60 mg/kg/week) vs. 69 placebo for 2–2.5 years	Significant 34% reduction in the annual lung density rate of decline measured by CT
Chapman, 2015 [39]	RAPID study extension: 140 cases treated with 60 mg/kg/week	Patients switched from placebo to therapy experienced a significant reduction in the emphysema progression

COPD, Chronic obstructive lung disease; CT, computed tomography; DLCO, diffusing capacity or transfer factor of the lung for carbon monoxide (CO); FEV1, forced expiratory volume in 1 s.

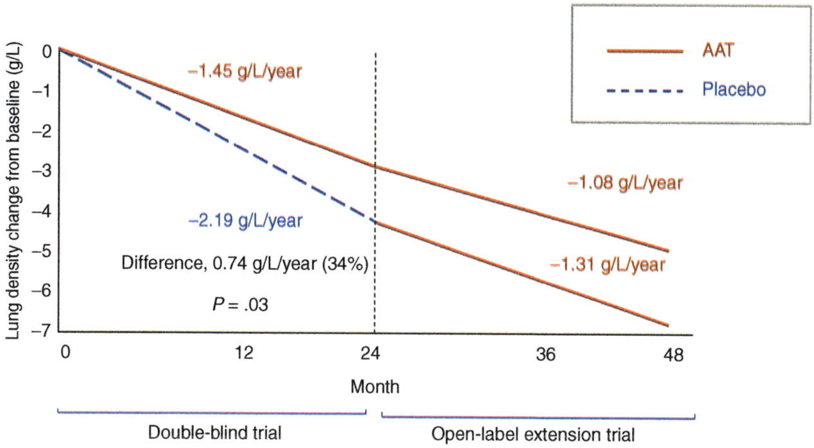

FIGURE 13.1 RAPID Trial: rates of lung density decrease at TLC during the double-blind and the open-label period. The *vertical dotted line* separates the first part of the study (the double-blind trial) of the open-label extension trial. Values on the graph are the annual rates of lung density decrease at total lung capacity (TLC). *(Adapted from Chapman KR, Burdon JG, Piitulainen E, Sandhaus RA, Seersholm N, Stocks JM, et al. Intravenous augmentation treatment and lung density in severe α1 antitrypsin deficiency (RAPID): a randomised, double-blind, placebo-controlled trial. Lancet 2015;386:360-8 [39].)*

The Canadian guide was updated in 2012 using a complicated GRADE (Grading of Recommendation Assessment, Development and Evaluation) methodology to analyze scientifically the relevant literature published between 1980 and 2011. This guide supports consideration of augmentation therapy in nonsmoking or ex-smoking patients with FEV_1 of 25%–80% predicted, attributable to emphysema and documented AATD (serum level ≤11 μM/L), who are receiving optimal pharmacological and nonpharmacological therapies (including pulmonary rehabilitation) because of benefits in CT scan lung density and mortality [42].

The Spanish Society of Pneumology and Thoracic Surgery guidelines 2015 support consideration of augmentation therapy in nonsmokers or ex-smokers, aged more than 18 years, with serum AAT concentrations <35% of normal, COPD defined by FEV_1/FVC <70% and FEV1 <80%, emphysema demonstrated by lung function test and X-ray computed tomography (CT), as well as in those showing accelerated loss of lung function, always in the absence of IgA deficiency [43].

Noteworthy, some experts believe that guidelines should be considered only as guidance, and the augmentation therapy indication should be individualized by combining a number of variables, such as age, comorbidities, physiological impairment, rate of lung function decline, rate of exacerbations, personal circumstances, and quality of life. In addition, according to these authors, individual pharmacokinetic studies may permit the selection of the best regimen of administration for those who need it [44].

Of note, the Medical and Scientific Advisory Committee of the Alpha-1 Foundation in 2008 argued against the use of augmentative therapy in MZ subjects concluding that clinicians should avoid prescribing augmentation therapy for this heterozygote population [45]. In addition, prophylactic treatment with intravenous AAT augmentation therapy for individuals with severe AATD, but no demonstrable emphysema, is not recommended [5].

Posttransplantation augmentation therapy has generally not been recommended. However, the ATS/ERS Task Force favors its use for lung transplant recipients during episodes that provoke inflammation and in situations of chronic rejection [1].

13.4.5 Available Drugs for Intravenous Augmentation Therapy

The currently approved products for replacement therapy are as follows: Prolastin, Prolastin C, and Trypsone (all three from Grifols), Aralast and Aralast NP (both from Baxter), Zemaira and Respreeza (both from CSL Behring), Alfalastin (from the Laboratoire français fractionnement et de Biotechnologies), and Glassia (from Kamada). All are presented in powder and solvent for intravenous infusion, except Glassia which is presented in liquid form. The high concentration after reconstitution provided by Prolastin C, Alfalastin, Respreeza, and Zemaira (i.e., 50 mg/mL) reduces infusion time to 15 min, and these data are very favorable when compared to those of the other preparations (i.e., Prolastin: 25 mg/ml and 30 min; Trypsone, Aralast, and Aralast NP: 20 mg/ml and 40 min; and Glaxia: 20 mg/ml and 60–80 min) [40,46] Table 13.2.

Based on pooled data provided by several studies [16,28,46,47], it has been concluded that infusions of Prolastin, Zemaira, and Aralast all provide serum levels above the protective threshold of 11 μM/L [48].

However, there are differences in purity and specific activity between these products, with a purity of 99% for Zemaira, compared with 70% for Aralast and 62% for Prolastin, which determines a higher specific activity of Zemaira compared with the other two drugs (Table 13.3), although the new preparations of Prolastin and Aralast (Prolastin C and Aralast NP) have improved both their purity and specific activity compared with the original products (92% and 0.7% vs. 62% and 0.6%, respectively). Impurities are due to the presence of other nontherapeutic proteins, including albumin, transferrin, haptoglobin, antithrombin III, antichymotrypsin, and IgA. In this sense, Zemaira clearly outperforms the rest in purity [28,48].

Another inequality between AAT products is given by the difference in their content of isoforms M2, M4, and M6. When exposed the AAT to the effect of a magnetic field with a pH range between 4 and 5, the protein melts in 9 bands called M0–M8. The most marked bands are 2, 4, and 6 (Fig. 13.2). Band 2 has one bi- and two tri-antennary side chains; band 4 has two bi- and one tri-antennary side chains, and band 6 has three bi-antennary side chains. Under these conditions, when comparing Aralast, Prolastin, and Zemaira with MM serum, it has been observed that all these commercial products have more M2 than normal serum, especially Aralast; Zemaira has more M4 than normal serum and the other products, which contain slightly lower concentration than normal serum.

TABLE 13.2 General Characteristics of Commercial Preparations of Alpha-1 Antitrypsin

Product trade name and pharmaceutical company	Sterilization method[a]	Pharmaceutical form of presentation	Concentration after reconstitution (mg/mL)	Infusion time (min)
Prolastin; Grifols (Spain)	Pasteurization	Powder and solvent for infusion	25	30
Prolastin C; Grifols (Spain)	Pasteurization + nanofiltration	Powder and solvent for infusion	50	15
Trypsone; Grifols (Spain)	S/D purification + nanofiltration	Powder and solvent for infusion	20	40
Aralast; Baxter (USA)	S/D purification + nanofiltration	Powder and solvent for infusion	20	40
Aralast NP; Baxter (USA)	S/D purification + nanofiltration	Powder and solvent for infusion	20	40
Zemaira; CSL Behring (USA)	Pasteurization + nanofiltration	Powder and solvent for infusion	50	15
Respreeza; CSL Behring (USA)	Pasteurization + nanofiltration	Powder and solvent for infusion	50	15
Alfalastin; LFB (France)	Pasteurization	Powder and solvent for infusion	50	15
Glaxia; Kamada (Israel)	Nanofiltration + S/D purification	Liquid	20	60–80

LFB, Laboratoire français de Fractionnement et de Biotechnologies; S/D, solvent/detergent.
[a]*All commercial AAT proteins are concentrated in the Cohn fraction IV of human plasma, which is prepared by cold ethanol fractionation.*

In addition, these three commercial products have lower M6 than normal serum. Of note, the increase of the total AAT concentration during inflammatory processes among normal subjects is primarily caused at the expense of M2 and M4 isoforms, and a possible consequence of the changed proportions can be the increased catabolism of the partially desialylated tri-antennary isoforms compared with that of the predominant bi-antennary form of the healthy individuals [49].

In conclusion, (1) all commercial preparations of AAT, at the doses recommended by the manufacturers, provide serum levels above the protective threshold. (2) Differences in purification techniques determine the purity and heterogeneity of the products. (3) Nevertheless, these differences in purity and

TABLE 13.3 Purity and Specific Activity[a] of Prolactin Aralast and Zemaira

	Prolastin	Aralast	Zemaira
Purity (%)	62	70	95
Specific activity	++	++	+++
Albumin	+	+	−
Transferrin	±	+	−
Haptoglobin	+	+	±
Antithrombin III	+	+	+
α1-Antichymotrypsin	+	+	+
Immunoglobulin G	+	+	±

[a]Specific activity is the activity of an enzyme per milligram of total protein (expressed in μmol/min/mg). Specific activity gives a measurement of enzyme purity in a mixture; ultra-high purity factors have a high specific activity.

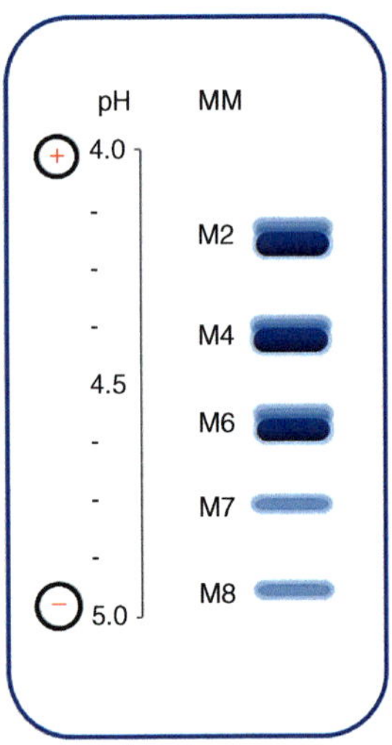

	M2	M4	M6
Aralast	31 (4)	40 (5)	20 (3)
Prolastin	13 (2)	39 (5)	48 (5)
Zemaira	13 (3)	48 (5)	38 (4)
Normal serum	8 (3)	41 (7)	54 (9)

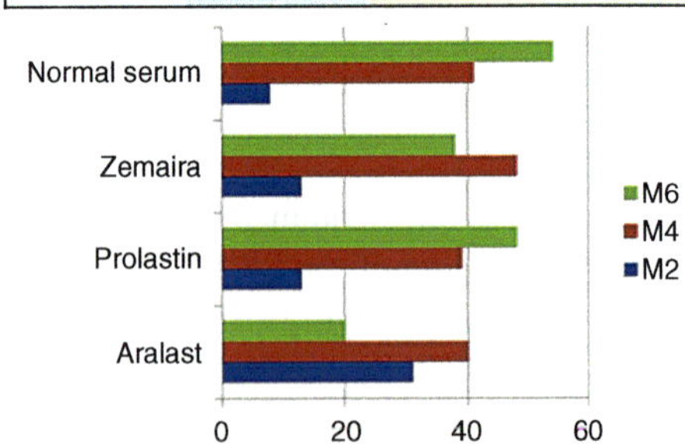

FIGURE 13.2 Aralast, Prolastin, and Zemaira M2, M4, and M6 isoforms contents compared with normal serum. On the left, main bands of the alpha-1 antitrypsin in isoelectric focusing. In the table the relative percentage of protein concentrations (standard deviation) of the M2, M4, and M6 isoforms are shown. Below, a comparative chart is displayed.

isoforms do not apparently modify the activity, efficacy, and safety of each commercial preparation, which are broadly considered as bioequivalent. (4) However, more studies are needed to determine with certainty whether these differences may impact on the clinical effectiveness and on the rate and severity of adverse effects.

13.4.6 Treatment Patterns

Regulatory agencies, national and international guidelines, and manufacturing laboratories recommend weekly doses of 60 mg/kg of AAT to treat patients with symptomatic AATD-related COPD [1]. However, long-term administration of weekly doses lifelong is not well accepted by some patients, and therefore other strategies have been suggested extending the interval between doses to 14, 21, or 28 days. For example, an alternate dose regimen of 250 mg/kg every 28 days maintained AAT levels above the protective threshold during the first 25 days, but fell below it during the last 3 days [20]. Remarkably, using pharmacokinetic simulations, it was found that protective serum AAT concentrations may be achieved with weekly administration of 60 mg/kg, and with biweekly administration of 120 mg/kg, but 180 mg/kg given every 21 days were inappropriate [50] (Fig. 13.3).

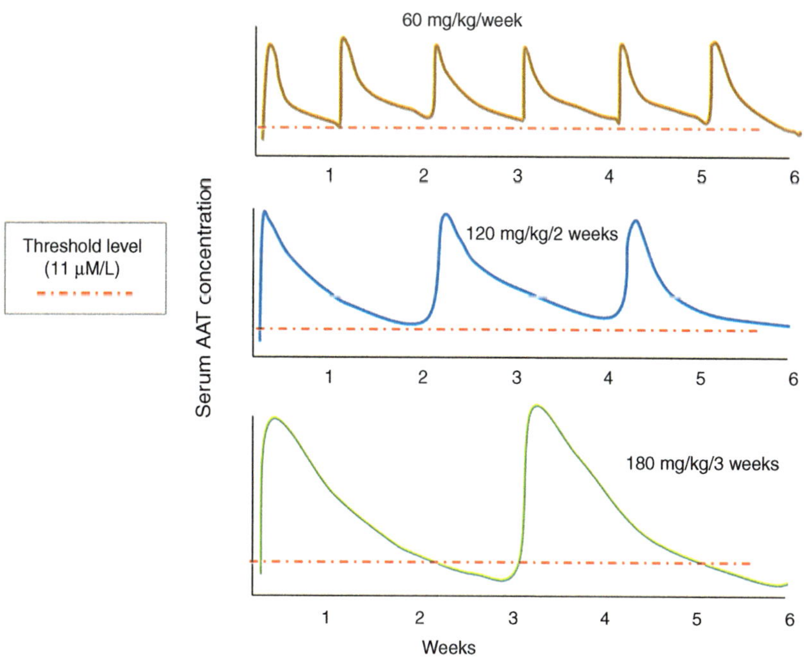

FIGURE 13.3 Simulated serum alpha-1 antitrypsin concentrations with different intravenous infusion schedules. The patterns 60 mg/kg/week and 120 mg/kg/2 weeks offer values above the protection threshold, whereas the pattern of 180 mg/kg/3 weeks leaves below the threshold of protection the last days before the next dose.

In clinical practice, patients tend to lengthen the time interval between infusions over time, as it has been observed in the NHLBI Registry, where at baseline 51% patients received AAT therapy once weekly, 26% every 2 or 3 weeks, and 21% once monthly, while at the end of the Registry follow-up, 33% received therapy every week, 43% every 2–3 weeks, and 24% once monthly [28].

13.4.7 Adverse Effects

Augmentative therapy is in general safe and well tolerated with few and minor adverse effects. In the largest database of adverse effects (the NHLBI Registry) the frequency of such effects was 0.02 per patient per month, but only 9% of these were considered serious and only 1.7% required emergency department attention or led to hospitalization. Up to 85% of patients reported no adverse effects. The most common adverse effects were headache (47%), dizziness (17%), nausea (9%), and dyspnea (9%). No adverse reactions to protein overload have been observed after long-term regular monthly administration of high doses of AAT. Patients who received weekly infusions reported a higher frequency of both adverse effects and adverse effects considered to be serious. To date, no occurrence of hepatitis, HIV, or prion disease transmission has been ascribed to this therapy [33,47,52].

REFERENCES

[1] American Thoracic Society/European Respiratory Society. American Thoracic Society/European Respiratory Society statement: standards for the diagnosis and management of individuals with alpha-1 antitrypsin deficiency. Am J Respir Crit Care Med 2003;168:818–900.

[2] Silverman EK, Sandhaus RA. Clinical practice. Alpha1-antitrypsin deficiency. N Engl J Med 2009;360:2749–57.

[3] Vestbo J, Hurd SS, Agustí AG, Jones PW, Vogelmeier C, Anzueto A, et al. Global strategy for the diagnosis, management, and prevention of chronic obstructive pulmonary disease: GOLD executive summary. Am J Respir Crit Care Med 2013;187:347–65.

[4] Stoller JK, Aboussouan LS. Alpha1-antitrypsin deficiency. Lancet 2005;365:2225–36.

[5] Stoller JK, Lacbawan FL, Aboussouan LS. Alpha-1 antitrypsin deficiency. In: Pagon RA, Adam MP, Ardinger HH, editors. GeneReviews® [Internet]. Seattle, WA: University of Washington; 1993–2016.

[6] Hogarth DK, Rachelefsky G. Screening and familial testing of patients for alpha 1-antitrypsin deficiency. Chest 2008;133:981–8.

[7] McGee D, Strange C, McClure R, Schwarz L, Erven M. The Alpha-1 Association Genetic Counseling Program: an innovative approach to service. J Genet Couns 2011;20:330–6.

[8] Hill AT, Campbell EJ, Bayley DL, Hill SL, Stockley RA. Evidence for excessive bronchial inflammation during an acute exacerbation of chronic obstructive pulmonary disease in patients with alpha-1-antitrypsin deficiency (PiZ). Am J Respir Crit Care Med 1999;160:1968–75.

[9] Whitson BA, Hayes D Jr. Indications and outcomes in adult lung transplantation. J Thorac Dis 2014;6:1018–23.

[10] Trulock EP, Christie JD, Edwards LB, Boucek MM, Aurora P, Taylor DO, et al. Registry of the International Society for Heart and Lung Transplantation: twenty-fourth official adult lung and heart-lung transplantation report—2007. J Heart Lung Transplant 2007;26:782–95.

[11] Stoller JK, Gildea TR, Ries AL, Meli YM, Karafa MT. National Emphysema Treatment Trial Research Group. Lung volume reduction surgery in patients with emphysema and alpha-1 antitrypsin deficiency. Ann Thorac Surg 2007;83:241.

[12] Donahue JM, Cassivi SD. Lung volume reduction surgery for patients with alpha-1 antitrypsin deficiency emphysema. Thorac Surg Clin 2009;19:201–8.

[13] Cohn EJ, Oncley JL, Strong LE, Hughes WL, Armstrong SH. Chemical, clinical, and immunological studies on the products of human plasma fractionation. I. The characterization of the protein fractions of human plasma. J Clin Invest 1944;23:417–32.

[14] Cohen H, Kernoff PB. ABC of transfusion. Plasma, plasma products, and indications for their use. BMJ 1990;300:803–6.

[15] Gadek JE, Hosea SW, Gelfand JA, Santaella M, Wickerhauser M, Triantaphyllopoulos DC, et al. Replacement therapy in hereditary angioedema: successful treatment of acute episodes of angioedema with partly purified C1 inhibitor. N Engl J Med 1980;302:542–6.

[16] Wewers MD, Crystal RG. Alpha-1 antitrypsin augmentation therapy. COPD 2013;10(Suppl. 1): 64–7.

[17] Gadek JE, Klein HG, Holland PV, Crystal RG. Replacement therapy of alpha 1-antitrypsin deficiency. Reversal of protease-antiprotease imbalance within the alveolar structures of PiZ subjects. J Clin Invest 1981;68:1158–65.

[18] Wewers MD, Casolaro MA, Sellers SE, Swayze SC, McPhaul KM, Wittes JT, et al. Replacement therapy for alpha 1-antitrypsin deficiency associated with emphysema. N Engl J Med 1987;316:1055–62.

[19] Schmidt EW, Rasche B, Ulmer WT, Konietzko N, Becker M, Fallise JP, et al. Replacement therapy for alpha-1-protease inhibitor deficiency in PiZ subjects with chronic obstructive lung disease. Am J Med 1988;84:63–9.

[20] Hubbard RC, Sellers S, Czerski D, Stephens L, Crystal RG. Biochemical efficacy and safety of monthly augmentation therapy for alpha 1-antitrypsin deficiency. JAMA 1988;260:1259–64.

[21] McElvaney NG, Stoller JK, Buist AS, Prakash UB, Brantly ML, Schluchter MD, et al. Baseline characteristics of enrollees in the National Heart, Lung and Blood Institute Registry of alpha 1-antitrypsin deficiency. Alpha 1-Antitrypsin Deficiency Registry Study Group. Chest 1997;111:394–403.

[22] Eden E, Hammel J, Rouhani FN, Brantly ML, Barker AF, Buist AS, et al. Asthma features in severe alpha-1 antitrypsin deficiency: experience of the National Heart, Lung, and Blood Institute Registry. Chest 2003;123:765–71.

[23] Crystal RG. Alpha-1-antitrypsin deficiency, emphysema and liver disease: genetic basis and strategies for therapy. J Clin Invest 1998;85:1343–52.

[24] Morse JO, Lebowitz MD, Knudson RJ, Burrows B. Relation of protease inhibitor phenotypes to obstructive lung diseases in a community. N Engl J Med 1977;296:1190–4.

[25] Hutchison DC, Tobin MJ, Cook PJ. Alpha 1 antitrypsin deficiency: clinical and physiological features in heterozygotes of Pi type SZ. A survey by the British Thoracic Association. Br J Dis Chest 1983;77:28–34.

[26] Turino GM, Barker AF, Brantly ML, Cohen AB, Connelly RP, Crystal RG, et al. Clinical features of individuals with PI*SZ phenotype of alpha 1-antitrypsin deficiency. Alpha 1-Antitrypsin Deficiency Registry Study Group. Am J Respir Crit Care Med 1996;154:1718–25.

[27] Green CE, Vayalapra S, Hampson JA, Mukherjee D, Stockley RA, Turner AM. PiSZ alpha-1 antitrypsin deficiency (AATD): pulmonary phenotype and prognosis relative to PiZZ AATD and PiMM COPD. Thorax 2015;70:939–45.

[28] Mohanka M, Khemasuwan D, Stoller JK. A review of augmentation therapy for alpha-1 antitrypsin deficiency. Expert Opin Biol Ther 2012;12:685–700.

[29] Teckman JH. Liver disease in alpha-1 antitrypsin deficiency: current understanding and future therapy. COPD 2013;10(Suppl. 1):35–43.

[30] Seersholm N, Wencker M, Banik N, Viskum K, Dirksen A, Kok-Jensen A, et al. Does alpha1-antitrypsin augmentation therapy slow the annual decline in FEV1 in patients with severe hereditary alpha1-antitrypsin deficiency? Wissenschaftliche Arbeitsgemeinschaft zur Therapie von Lungenerkrankungen (WATL) alpha1-AT study group. Eur Respir J 1997;10:2260–3.

[31] The Alpha-l-Antitrypsin Deficiency Registry Study Group. Survival and FEVl decline in individuals with severe deficiency of alpha-l antitrypsin. Am J Respir Crit Care Med 1998; 158:49–59.

[32] Wencker M, Fuhrmann B, Banik N, Konietzko N. for Wissenschaftliche Arbeitsgemeinschaft zur Therapie von Lungenerkrankungen. Longitudinal follow-up of patients with alpha(1)-protease inhibitor defi ciency before and during therapy with IV alpha(1)-protease inhibitor. Chest 2001;119:737–44.

[33] Tonelli AR, Rouhani F, Li N, Schreck P, Brantly ML. Alpha-1-antitrypsin augmentation therapy in deficient individuals enrolled in the Alpha-1 Foundation DNA and Tissue Bank. Int J Chron Obstruct Pulmon Dis 2009;4:443–52.

[34] Chapman KR, Stockley RA, Dawkins C, Wilkes MM, Navickis RJ. Augmentation therapy for alpha1 antitrypsin deficiency: a meta-analysis. COPD 2009;6:177–84.

[35] Carlson MD, Morrison RS. Study design, precision, and validity in observational studies. J Palliat Med 2009;12:77–82.

[36] Burrows B. A clinical trial of efficacy of antiproteolytic therapy: can it be done? Am Rev Respir Dis 1983;127:S42–3.

[37] Dirksen A, Dijkman JH, Madsen F, Stoel B, Hutchison DC, Ulrik CS, et al. A randomized clinical trial of alpha(1)-antitrypsin augmentation therapy. Am J Respir Crit Care Med 1999;160:1468–72.

[38] Dirksen A, Piitulainen E, Parr DG, Deng C, Wencker M, Shaker SB, et al. Exploring the role of CT densitometry: a randomised study of augmentation therapy in alpha1-antitrypsin deficiency. Eur Respir J 2009;33:1345–53.

[39] Chapman KR, Burdon JG, Piitulainen E, Sandhaus RA, Seersholm N, Stocks JM, et al. Intravenous augmentation treatment and lung density in severe α1 antitrypsin deficiency (RAPID): a randomised, double-blind, placebo-controlled trial. Lancet 2015;386:360–8.

[40] Greulich T, Vogelmeier CF. Alpha-1-antitrypsin deficiency: increasing awareness and improving diagnosis. Ther Adv Respir Dis 2016;10:72–84.

[41] Guidelines for the approach to the patient with severe hereditary alpha-1-antitrypsin deficiency. American Thoracic Society. Am Rev Respir Dis. 1989;140:1494–7.

[42] Marciniuk DD, Hernandez P, Balter M, Bourbeau J, Chapman KR, Ford GT, et al. Alpha-1 antitrypsin deficiency targeted testing and augmentation therapy: a Canadian Thoracic Society clinical practice guideline. Can Respir J 2012;19:109–16.

[43] Casas F, Blanco I, Martínez MT, Bustamante A, Miravitlles M, Cadenas S, et al. Indications for active case searches and intravenous alpha-1 antitrypsin treatment for patients with alpha-1 antitrypsin deficiency chronic pulmonary obstructive disease: an update. Arch Bronconeumol 2015;51:185–92.

[44] Stockley RA, Miravitlles M, Vogelmeier C. Alpha One International Registry (A.I.R.). Augmentation therapy for alpha-1 antitrypsin deficiency: towards a personalised approach. Orphanet J Rare Dis 2013;8:149.

[45] Sandhaus RA, Turino G, Stocks J, Strange C, Trapnell BC, Silverman EK, et al. alpha1-Antitrypsin augmentation therapy for PI*MZ heterozygotes: a cautionary note. Chest 2008;134:831–4.

[46] Stoller JK, Rouhani F, Brantly M, Shahin S, Dweik RA, Stocks JM, et al. Biochemical efficacy and safety of a new pooled human plasma alpha(1)-antitrypsin, Respitin. Chest 2002;122:66–74.

[47] Stocks JM, Brantly M, Pollock D, Barker A, Kueppers F, Strange C, et al. Multi-center study: the biochemical efficacy, safety and tolerability of a new alpha1-proteinase inhibitor, Zemaira. COPD 2006;3:17–23.

[48] Kueppers F. The role of augmentation therapy in alpha-1 antitrypsin deficiency. Curr Med Res Opin 2011;27:579–88.

[49] Vaughan L, Lorier MA, Carrell RW. Alpha 1-antitrypsin microheterogeneity. Isolation and physiological significance of isoforms. Biochim Biophys Acta 1982;701:339–45.

[50] Soy D, de la Roza C, Lara B, Esquinas C, Torres A, Miravitlles M. Alpha-1-antitrypsin deficiency: optimal therapeutic regimen based on population pharmacokinetics. Thorax 2006;61:1059–64.

[51] Stoller JK, Fallat R, Schluchter MD, O'Brien RG, Connor JT, Gross N, et al. Augmentation therapy with alpha1-antitrypsin: patterns of use and adverse events. Chest 2003;123:1425–34.

[52] Stoller JK, Fallat R, Schluchter MD, O'Brien RG, Connor JT, Gross N, O'Neil K, Sandhaus R, Crystal RG. Augmentation therapy with alpha1-antitrypsin: patterns of use and adverse events. Chest 2003;123:1425–34.

Chapter 14

Registries and Patients' Associations

14.1 INTRODUCTION

Severe alpha-1 antitrypsin (AAT) deficiency (AATD) is an uncommon condition, typically linked to genotypes Pi*Z, that has been included among the rare diseases group. At present, there is not a universal definition for rare disease. For example, a disease or disorder is defined as rare in Europe when it affects fewer than 1 in 2000 people, or approximately 1 in 1500 people in the United States. Nevertheless, despite the discrepancies in terminology and prevalence used to define a rare disease, an average prevalence threshold between 40 and 50 cases/100,000 people is reasonably accepted [1].

Severe Pi*Z-AATD is an inherited condition that mainly affects Caucasian individuals, having its highest prevalence in Northern and Western Europe (1:2000–1:2500 individuals). In Central Europe, USA, Canada, Australia, and New Zealand, its prevalence is around 1:4000–1:5000. Pi*Z prevalence declines significantly in Latin America, and is rare or inexistent in no-Caucasian people from African and Asian countries [2].

Broadly, it has been estimated that it might be approximately 250,000 Pi*ZZ worldwide (roughly distributed as follows: 120,000 in Europe; 74,000 in North America—85% of them, non-Hispanic white Americans-; 3,000 in Central America; 15,000 in South America; 3,000 in Africa; 30,000 in Asia, 4,000 in Australia and 2,000 in New Zealand), and potentially up to 60% of them may develop COPD, approximately 30% may be affected by liver cirrhosis, 2% may develop systemic vasculitis, almost 1% may suffer from neutrophilic panniculitis, and a minimum percentage of these individuals may experience several of these diseases at the same time [3].

Even though the condition was identified for the first time in 1963, more than five decades later its low frequency and underdiagnoses are still major handicaps to properly understand and manage it. In addition, the marked variability of the AATD manifestations hinders knowing properly the natural history and prognosis of the different diseases with which Pi*Z-AATD can manifest, and its prevention and cure are still far away [4].

In this context, registries of patients with AATD constitute key instruments to develop clinical and basic research, to evaluate the safety and efficacy of available therapies, to promote drugs development and clinical trials, to connect patients and families affected by this disorder, and to improve patient care and health care planning [5].

14.2 THE ALPHA ONE INTERNATIONAL REGISTRY

In March 1996, a World Health Organization meeting recommended the establishment of national and international registries to facilitate data collection and collaborative research, as well as helping to create a patient resource bank for the design and conduct powered clinical trials [6].

To comply with these recommendations, several key opinion leaders in Europe established the Alpha One International Registry (AIR), initially founded by uniting the different existing European national registries. The main initial objective was to constitute a comprehensive database of AATD patients to increase the understanding of the condition. Other proposed objectives were to promote collaborative basic and clinical research, to facilitate clinical trials development, to disseminate information about the AATD-related diseases, and to encourage support and awareness of AATD among people responsible for health [7].

Currently, AIR is a web-based multinational research organization, located in Malmo (Sweden), representing a combined research effort of 18 countries, including United Kingdom, Germany, Sweden, Denmark, Finland, Poland, Belgium, Holland, Austria, Switzerland, Italy, Spain, Israel, Canada, Argentina, Brazil, Australia, and New Zealand. A simple but extensive web-enabled questionnaire has been developed for AIR, which may be completed online by authorized physicians from all these countries. Each representative of national registries is in charge of passing the data of the patients to the international database and to update them regularly. The quality of data provided by the national registries is audited at the central AIR database by a manager, who supervises all national entries and the patient's follow-up. The information about patients is anonymous and absolutely confidential; patients are listed with the initials of his name and an identification number, so that only the physician responsible for a patient knows these identification data [8].

In a report of 2013, AIR included 4758 patients, with the following Pi* genotypes (numbers in parentheses): ZZ (4031), SZ (538), Mmalton/Z (19), Null (17), IZ (12), and FZ (4) [9]. The number of volunteers with AATD has increased progressively, and on November 2014 the AIR included 5147 subjects (Fig. 14.1).

14.3 EUROPEAN NATIONAL REGISTRIES

14.3.1 Austria

The Austrian National Registry for individuals with AATD was founded on 1996 to get new insights into the pathogenesis of the disease and to develop therapeutic strategies. According to a 2010 report, the registry had then enrolled 139 severely affected AATD patients, most of them smokers or ex-smokers (72%), who

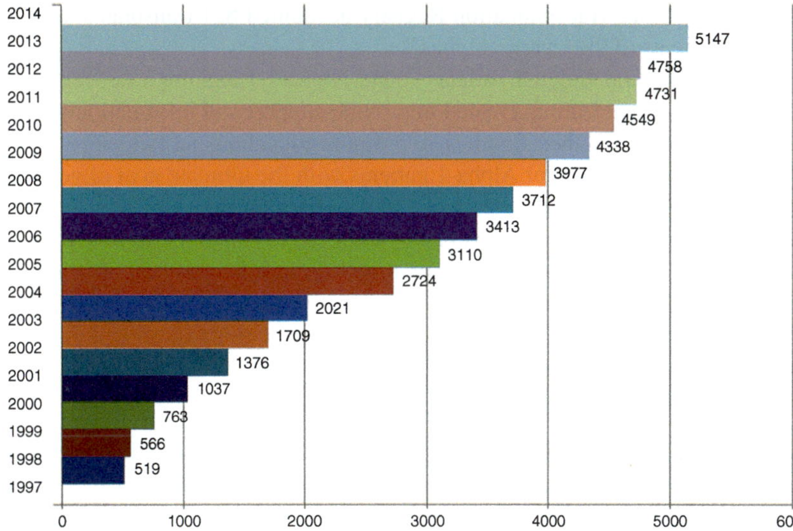

FIGURE 14.1 **Annual accumulation of patient numbers in AIR database (1997–2014).**

developed respiratory symptoms in their fourth decade. The mean duration between the onset of symptoms and the final diagnosis was 6.5 years. Approximately 25% of the evaluated patients were unable to practice their habitual profession because of the disability caused by the disease [10]. Currently, the Austrian Registry is member of AIR, and transfers the data of its cases to the international registry.

14.3.2 Belgium

In 1999, Belgium joined AIR, which resulted in 2001, in a first Belgian working group settled by the Belgian Thoracic Society. In 2003, this group estab lished the Belgian Alpha-1-antitrypsin Deficiency Registry. The objectives of the Belgian Registry were similar to those of most of the remaining registries, that is, to disseminate information concerning AATD, to encourage preventive measures, and to promote diagnostic screening and registration of cases.

Each Belgian chest physician enrolled in the Belgian Thoracic Society could have a password with access to the web log of the registry. The first Belgian patients were included in the AIR in 2003. After introduction of the patient's requested data, the local registry administrator validated the case if the data were correctly entered. In 2014, 55 patients (36 males and 19 females) were included. Forty-six of these patients were also included in the AIR. Only 9 were nonsmokers, but 36 had stopped smoking. Practically all patients with AATD suffered from lung disease; six were diagnosed in a context of family screening, and two had a history of neonatal hepatitis. Eighteen patients were lung transplanted, and 1 was liver transplanted. Sixteen patients had, on registration, benefited from augmentation therapy, and 5 from long-term oxygen therapy.

The Belgian Lung Transplantation Registry accounted 821 transplants, and 32 of these patients (3.9%) were "Alphas."

In 2012, several Belgian patients found Alpha-1 Belgium, a not-for-profit patient-driven organization, created to provide support and information for patients with AATD and their families, to promote education and awareness of the disease, and to advocate for Alpha-1 patients with the ultimate goal of generating awareness about the disease and to contact the Public Health Structures. In addition, in June 2014, the Belgian association created another important international structure called Alpha-1 Global, whose mission is to create a global network of Alpha-1 organizations and patients to increase awareness, cases detection, and access to care for "Alphas" around the world [11].

14.3.3 Denmark

Since 1979 Danish AATD patients have been notified to the national registry by pulmonologists and pediatricians from all over Denmark, and more than 2500 family members of the probands were tested then. As a result, in 1994 the registry consisted of 312 families, with 513 Pi*SZ and Pi*ZZ individuals [12]. The Danish Alpha-1 Antitrypsin Deficiency Registry is member of AIR since the creation of the international registry.

14.3.4 France

The French National Registry of Patients with AATD-related emphysema was launched in 2006, and has currently gathered information for approximately 300 patients. The aim of this registry is to better understand the natural history of the pulmonary disease related to the deficiency and to search for associated factors. The French registry includes many patients with severe respiratory impairment, with almost 70% of registered patients having a FEV_1 <50% predicted. At the time of inclusion in the registry, approximately 50% of registered patients were receiving augmentation therapy, and 32% were on long-term home oxygen therapy [13].

14.3.5 Germany

The German National Registry for Individuals with AATD was set up in the 1980s with the aim of getting information about the natural history and the quality of care for these patients. The German Registry is one of the founding members of the AIR. Since 2004 it is maintained in Marburg and Homburg.

In 2006, the German Registry organized a nation-wide case-detection program, identifying 335 patients with severe AATD including 16 individuals with rare genotypes [14]. In 2008, 548 individuals were registered [15]. In 2015, the number of subjects enrolled rose to 1066, with the following Pi genotypes: 820 Pi*ZZ (males, 56%), 109 Pi*SZ (males, 55%), and 137 rare deficiency types. The age of the registered individuals was 52 ± 13 years [16].

14.3.6 Ireland

The establishment of the Irish National AATD Targeted Detection Programme in 2004 increased the detection rate of AATD, and in 2007, the Irish National AAT Registry was created. The registry is maintained by the Irish Alpha One Foundation (AOF). The national referral center for AATD in Ireland is based on the Beaumont Hospital, Dublin [17].

The detection program has detected 250 Pi*ZZ and 185 Pi*SZ individuals, as well as a number of subjects with rare mutations, including a Null/Null homozygote and a Mmalton/Mmalton homozygote [18].

14.3.7 Italy

In 1993, a national study group on AATD was fostered by the Italian Pneumological Societies. Almost simultaneously, Balbi et al. reported the first case of a Pi*ZZ Italian patient with bronchiectasis [19]. Thereafter, a National Reference Center was launched at the Institute for Respiratory Disorders of the University of Pavia, where Maurizio Luisetti was in charge of this important task. The Italian Registry for AATD was established in 1996. Inclusion criteria included the presence of severe AATD, defined by the carriage of the Pi*ZZ genotype, Pi*SZ genotype, or another rare severe deficient or null genotypic variants. The patients' data were collected on forms sent by post or email to the Italian Centre for AATD diagnosis in Pavia, where the coordinators kept continuously updated the database. Informed consent prior to enrolment was obtained from all patients [20].

In a period spanning 18 years (1996–2014), a total of 422 adult subjects with AATD were enrolled, namely 258 Pi*ZZ, 74 Pi*SZ, 4 Pi*SS, and 86 patients with at least one rare or null deficient allele. The 21.3% frequency for AATD patients with at least one deficient rare variant is the highest so far recorded in national registries of AATD [21]. The Italian AATD registry became part of the AIR in 1999.

14.3.8 The Netherlands

The Netherlands started collecting baseline and prospective data right from the beginning of the web-based AIR patient database in 1998 [7]. Recorded data were obtained from local physicians or from patient visiting the Leiden University Medical Center.

In 2014, a total of 450 genotyped ZZ and 35 null variants, including Heerlen, Bredevoort, Bellingham, and Granite Falls, were registered. The Dutch Registry has mainly been used to recruit patients for randomized placebo-controlled studies with new treatment regimens [22], as well as for assessing the value of new biomarkers for severe AATD associated with pulmonary emphysema [23].

14.3.9 Poland

The first Polish AAT laboratory was set up in 2009 at the National Institute of Tuberculosis and Lung Diseases in Warsaw. Subsequently, in 2010 an AATD screening program targeting patients with respiratory disorders, together with a National Registry, was established in Poland. In parallel, this same year the Polish Respiratory Society published specific recommendations on AATD diagnostics and treatment, joining efforts of specialists in respiratory and liver disorders [24]. As augmentation therapy is currently not reimbursed in Poland, the early detection of cases to implement antismoking measures and optimal pharmacotherapy for COPD is essential in this country.

The registry resources include a principal coordinator, a consultant committee of six respiratory specialists and a computer support staff. The registry complies with all existing data protection legislation and has been approved by the competent Polish authorities. It is financially supported by the National Institute of Tuberculosis and Lung Diseases as well as by the Polish Foundation of Patients with AATD.

Between 2010 and 2014, a total of 2525 samples from respiratory patients were collected countrywide; 55 patients with severe AATD or rare mutations were identified and registered, including 36 Pi*ZZ subjects (65%). The majority of AATD patients were diagnosed not only with COPD (40%) or emphysema (7%), but also with bronchial asthma (16%) and bronchiectasis (13%). The average age of ZZ subjects at diagnosis was 49 ± 14 years. More than a half (56%) of AATD subjects recorded in the database have a history of smoking [25].

14.3.10 Spain

The first two cases of emphysema due to AATD receiving augmentation therapy with intravenous AAT in Spain were reported in 1991 [26]. Following the description of these initial cases, a growing interest in the deficiency resulted in the development of the Spanish Registry of Patients with Alpha-1 Antitrypsin Deficiency (REDAAT), which was founded in 1993 as a task force of the Spanish Society of Respiratory Medicine and Thoracic Surgery. The aim of the REDAAT has always been to increase the scientific knowledge about the deficiency through basic and clinical research, to raise awareness about the deficiency, and to provide support for the diagnosis through phenotyping and genotyping. The REDAAT recruited individuals mainly diagnosed not only with phenotypes Pi*ZZ, but also with Pi*SZ and carriers of rare and null deficiency variants. Follow-up information is requested every 6 months until death or transplantation.

REDAAT participated in the founding of the AIR, and the Spanish dataset was accordingly modified to be compatible with the AIR database, in order to the Spanish data were regularly uploaded to the AIR. To improve the quality of the collected data, upon notification of a new case an automatic alert is sent to the website manager who reviews and validates the information before it is definitively recorded. This system allows the manager to correct typing errors and to contact the referring physician if relevant information is missing or even to delete inappropriate registrations (e.g., nondeficient phenotypes or patients with incomplete or nonconfirmed diagnoses). REDAAT is also linked with the Spanish Rare Diseases Registry.

At present, the REDAAT is a network of health care professionals with an interest in AATD and consists of more than 360 physicians who registered patients from all communities in Spain. The website and database are directed by an advisory board including 12 pulmonologists, 3 pediatricians, and 3 basic researchers. Apart from the database, the website includes educational materials for patients and guidelines for diagnosis and treatment of individuals with AATD [27,28].

In January 2014, the REDAAT database included a total of 511 individuals (in June 2016, this number has risen to 649). The adult population consisted of 469 individuals (92% of the total) and their phenotypic distribution was 348 Pi*ZZ (74%), 100 Pi*SZ (21%), and 21 carriers of rare variants (5%). The most frequent diagnosis is lung disease (75%). A total of 61 deaths were recorded, and 45% percent of individuals received augmentation therapy [29].

14.3.11 Sweden

One of the major registries in Europe for AATD (along with the German and the UK registries) is located in Sweden, where the condition was first recognized in 1963 [30,31].

The Swedish National Register of Severe AATD was established in 1991 with the aim to prospectively study the natural history of the condition. The main inclusion criteria in the register were age (adults $\geq$18 years) and Pi*Z phenotype. The register is kept updated by means of repeated questionnaires providing data to allow analysis of the mode of identification, lung and liver function, smoking habits, respiratory symptoms, and diagnoses as reported by physicians. Until 2014, a total of 1553 Pi*ZZ individuals were included in the register. Forty-three percent of the subjects were identified by respiratory symptoms, 7% by liver disease, 26% by other pathological conditions, and 24% by population or family screening. Forty-five percent of the subjects had never smoked, 47% were ex-smokers, and 8% current smokers. Twenty-eight percent of the never-smokers, 72% of the ex-smokers, and 61% of the current smokers fulfilled criteria for COPD. Eleven hundred and two subjects are still alive, and among the 451 deceased the most common cause of death was respiratory disease (55%), followed by liver disease (13%) [32]. The Swedish National Register is part of AIR since the creation of the international registry.

14.3.12 United Kingdom

The Antitrypsin Deficiency Assessment and Programme for Treatment (ADAPT) Registry was established in 1996 to enable patients to be seen from all regions of the United Kingdom with the purpose of collecting extensive demographics. Recruitment started in 1997, and in 2014, the ADAPT Registry had 1065 patients on the database, that is, 930 Pi*ZZ and 135 Pi*SZ. Patients had been seen generally once a year, at which time full demographics (including health status questionnaires), full post bronchodilator lung function testing, CT scanning as indicated clinically or for research purposes, routine bloods for hematology and liver function, research bloods for potential biomarkers, whole blood for DNA, sputum for quantitative culture, and diary cards for monitoring exacerbations were collected. Other investigations, such as CT follow-up of abnormalities, liver ultrasound, and biopsies were carried out when indicated, in collaboration with the patient's local medical services. The ADAPT model has proven to be highly productive providing new insights especially into the lung disease, generating and delivering clinical trials and establishing active patient groups and participation.

Of note, it has been the long-term commitment of the patients who have attended on an annual basis from all parts of the United Kingdom which have facilitated the development of an active patient group closely allied to the ADAPT Registry. The collaboration between clinicians, researchers, and patients generates strong patient support groups, such as the Alpha-1 Alliance UK. These groups ensure continued understanding of patient impact and aims, as well as strong support for other affected individuals [33].

14.4 AMERICAN REGISTRIES

14.4.1 Canada

The Canadian Alpha-1 Antitrypsin Deficiency Registry was formed in 1999 as part of an international effort to facilitate research in this uncommon disorder by creating a database of AATD individuals [7]. Since 1999, as part of the AIR, the Canadian Alpha-1 Antitrypsin Deficiency Registry has included patients with ZZ and other severe deficiency genotypes. All registry participants provided written informed consent to share their data with the registry. Two hundred and ninety patients (146 males), most Pi*ZZ, were enrolled by 2013. Typical patients were ex-smokers with respiratory symptoms, severely reduced FEV_1, accelerated rate of FEV_1 decline, and minimal bronchodilator response. Patients with lung disease reported symptoms onset at 40 ± 11 years, but were diagnosed as AATD at 47 ± 10 years. A subgroup diagnosed by liver disease or familial screening was younger and had better preserved lung function. There were 63 deaths, of which 29 were lung-related and 6 liver-related. Average age at death was 60 ± 11 years. There was a marked regional variation in the rate of diagnosis and augmentation therapy usage [34].

14.4.2 The United States

14.4.2.1 The National Heart, Lung and Blood Institute Alpha-1-Antitrypsin Deficiency Registry

Alpha-1-Antitrypsin Deficiency recognition increased dramatically in the late 1980s, following the introduction of augmentation therapy in the United States. An exemplary collaboration between Cutter Biological (a subsidiary of Miles Laboratories), the Food and Drug Administration (FDA), and the National Heart, Lung and Blood Institute (NHLBI) of the National Institutes of Health (NIH) formed to develop a prospective registry of individuals on or off augmentation therapy. Thereby, from 1989 to 1992 the NHLBI Registry enrolled individuals with severe AATD on or off augmentation therapy, and followed them until 1996 [35]. The resulting dataset of 1129 individuals allowed a much better understanding of the natural history of AATD.

When the NHLBI Registry ended, a community of patients and health care professionals committed to better understand AATD leading to the formation of the Alpha-1 Foundation (AOF), AlphaNet, and the Alpha-1 Foundation Research Registry.

14.4.2.2 The Alpha-1 Foundation Research Registry

The AOF, a patient-organized research foundation, is a not-for-profit corporation founded in 1995 by John Walsh, Sandy Lindsey, and Susan Stanley, three people diagnosed with AATD (colloquially "Alpha-1"). A majority of the AOF Board of Directors is either diagnosed with Alpha-1 or has a family member diagnosed with Alpha-1.

The AOF has developed a solid infrastructure to promote research and development of new therapies for improving the quality of life for those diagnosed with Alpha-1. It has fostered collaborations with investigators throughout the United States and Europe, working closely with the NIH, the FDA, individuals affected by Alpha-1, and the pharmaceutical industry to expedite the development of improved therapies. Using words of their own leaders, the mission of the AOF is "to finding a cure for AATD and to improving the lives of people affected by Alpha-1 worldwide."

The AOF has collaborated with clinician–scientists to organize the Alpha One Foundation Research Network Registry, with the purpose of facilitating clinical research in AATD. Elements of the Alpha One Foundation Research Network Registry included a Medical and Scientific Advisory Committee, composed of physician–investigators and patient advocates, designated clinical resource centers at medical institutions with expertise in the management of individuals with AATD, and a data coordinating center with responsibility for database management and analysis. Questionnaires requesting information about demographic features, AAT pheno/genotype, smoking history, and health care utilization have just been distributed to prospective registrants through the following channels: mailings from the AOF; mailings from the clinical resource centers; and distribution by home-care and pharmaceutical companies. According to a report published 3 years after its founding, 712 volunteers were registered between 1997 and 1999. The characteristics of the registered volunteers were as follows: mean age, 49 ± 13 years; women, 48%; white, 96%; Pi*Z, 71%; ex-smokers, 73%; COPD, 87%; liver disease, 6%; treated with IV augmentation therapy, 59%; and receiving supplemental oxygen therapy at home, 35% [36].

The Alpha-1 Foundation Research Registry is being used to invite participants for institutional review board (IRB)-approved research studies in the United States and Canada. The Registry design allows enrolment online to include everyone with a deficient AAT allele. Stratification by genotype, geography, or self-reported lung or liver disease allows targeted invitations by letters approved by an IRB. Incentives for participation, travel monies, or study specific testing schedules are usually hosted by the study Principal Investigator.

In 2001, individuals who desired confidential family testing with accurate diagnostic tests signed electronic consent with the Alpha-1 Foundation Research Registry, answered questionnaires concerning the reasons and projected impact of testing, and were mailed a blood spot card, lancet, and instructions for blood spot acquisition. Blood spot cards were mailed to the Registry, anonymized, and tested at the University of Florida Alpha-1 laboratory, with results returned to the Medical University of South Carolina. Results were later returned to the individuals with specific information, access to a genetic counselor for themselves and their family members, and specific invitations to join the Registry if they were AATD. Enrolment of mildly deficient genotypes began in 2002, anticipating the increased interest in carriers of one severely deficient AAT gene.

This program has the highest yield of diagnostic testing yet reported. The Alpha-1 Foundation Research Registry continues to enroll at approximately 300 individuals yearly from every state in the United States and through an enrolment agreement with every province in Canada [37].

14.5 PATIENT CARE STRUCTURES

14.5.1 AlphaNet

In response to the wide variability of care in the United States, patients themselves developed a not-for-profit organization to provide disease management service to the AATD patients. This organization, AlphaNet, is a unique patient-driven business model that pushes advocacy through customized patient care and disease management while funding research for a cure.

AlphaNet has developed the Alpha-1 Disease Management and Prevention Program (ADMAPP), consisting in a comprehensive written educational patient-directed material for self-study and treatment plans. The program has provided patient self-management of AATD to more than 3000 individuals. The

TABLE 14.1 Links and Addresses of Some of the Main Registries and Associations of Patients With Alpha-1 Antitrypsin Deficiency

Alpha-1 Foundation	www.alpha1.org	3300 Ponce de Leon Blvd. Coral Gables, Florida 33134; Main: (305) 567-9888; Toll Free: (877) 2-CURE-A1 or (877) 228-7321; Fax: (305) 567-1317
Alpha-1 Foundation Research Registry	https://alphaoneregistry.org/	3300 Ponce de Leon Blvd. Coral Gables, FL 33134; Phone: (877) 228-7321; Email: info@alphaone.org
Alpha One International Registry (AIR)	www.aatregistry.org/	–
Alpha-1 Association	www.alpha1.org	815 Connecticut Ave., NW, Suite 1200, Washington, DC 20006-4004; Phone: (202) 887-1900; Toll Free: (800) 521-3025; Fax (202) 887-1964; Email: info@alpha1.org
Alpha-1 Global	www.alpha-1global.org/	Global: Gonny Gutierrez, Director; Office: +1 305 567 9888, extension 212; Mobile: +1 303 356 5540; Email: ggutierrez@alpha-1global.org Skype: gonny.gutierrez 3300 Ponce de Leon Boulevard, Miami, FL 33134, USA Europe: Alpha-1 Global Steering Committee Chair, Frank Willersinn, MD, President Alpha-1 Plus; Office: +32 22 03 41 31; Mobile: +32 475 691 782; Email: alpha1plus@yahoo.be Skype: Sinopi77 Rue Bâtonnier Braffort/Stafhouder Braffortstraat 58, 1200 Bruxelles/Brussels, Belgium Australia and Asia: Alpha-1 Global Liaison, Steven Knowles, President Alpha-1 Association of Australia; Office: +61 7 3103 3363; Mobile: +61 410 108 104; Email: steven.knowles@alpha1.org.au Skype: steven-knowles GPO Box 2100, Brisbane QLD 4001, Australia

program includes access to the Big Fat Reference Guide to Alpha-1 (BFRG) and assignment to an AlphaNet Disease Management Coordinator (ANC). ANCs are individuals with AATD, specially trained in the ADMAPP program. Each follows 100–150 patients by telephone at least once monthly. In addition to answering questions from their patients, they administer health and quality of life questionnaires, and help them in many other different aspects related to the disease and other aspects of daily living. The effectiveness of this program has been demonstrated by a prospective study showing that ADMAPP improved the health outcomes in 878 subjects with AATD-associated COPD [38].

14.5.2 Alpha-1 Global

In 2012, several Belgian AATD patients decided to found Alpha-1 Belgium to create awareness about the disease and to contact the Public Health Structures, and soon after (in June 2014), this Belgian association created another new international structure called Alpha-1 Global. Alpha-1 Global is formed by a group of seven volunteer patients, who are representatives of Alpha-1 patient groups from Australia, Belgium, Brazil, Canada, Denmark, Portugal, Spain, USA, and the United Kingdom. Based on that "Alphas" worldwide experience many of the same following problems: little to no awareness of their condition; late diagnosis; physicians unfamiliar with alpha-1 care standards; limited or no access to augmentation therapy, and so on; the mission assigned to Alpha-1 Global is to create a collaborative network of Alpha-1 organizations and patients to increase awareness, detection, and access to care for Alphas around the world. Alpha-1 Global has set up an Internet platform for patients, medical and care professionals, and researchers to communicate about the disease [11].

The names and addresses of various relevant associations of patients with AATD are listed in Table 14.1.

REFERENCES

[1] Richter T, Nestler-Parr S, Babela R, Khan ZM, Tesoro T, Molsen E, et al. Rare disease terminology and definitions—a systematic global review: report of the ISPOR Rare Disease Special Interest Group. Value Health 2015;18:906–14.

[2] de Serres F, Blanco I. Role of alpha-1 antitrypsin in human health and disease. J Intern Med 2014;276:311–35.

[3] de Serres FJ, Blanco I. Prevalence of α1-antitrypsin deficiency alleles PI*S and PI*Z worldwide and effective screening for each of the five phenotypic classes PI*MS, PI*MZ, PI*SS, PI*SZ, and PI*ZZ: a comprehensive review. Ther Adv Respir Dis 2012;6:277–95.

[4] Miravitlles M. COPD: Journal of Chronic Obstructive Pulmonary Disease. Preface. COPD 2015;12(Suppl. 1):1.

[5] Miravitlles M, Brusasco V. Alpha1-antitrypsin deficiency: the contribution of national registries to the understanding of the disorder. A journal Supplement in memory of Professor Maurizio Luisetti. COPD 2015;12:1–2.

[6] Alpha1-antitrypsin deficiency: memorandum from a WHO meeting. Bull World Health Organ 1997;75:397–415.

[7] Stockley RA, Luisetti M, Miravitlles M, Piitulainen E, Fernandez P. Ongoing research in Europe: Alpha One International Registry (AIR) objectives and development. Eur Respir J 2007;29:582–6.

[8] Luisetti M. Introduction: the history of the Alpha One International Registry (A.I. R.). Respir Med 2000;94(Suppl. C):S1–2.

[9] Stockley RA, Dirksen A, Stolk J. Alpha-1 antitrypsin deficiency: the European experience. COPD 2013;10(Suppl. 1):50–3.

[10] Huber F, Schmid-Scherzer K, Wantke F, Frantal S, Kneussl M. [Alpha1-antitrypsin deficiency in Austria: analysis of the Austrian Alpha1-international-registry database] [Article in German]. Wien Klin Wochenschr 2010;122:390–6.

[11] Hutsebaut J, Janssens W, Louis R, Willersinn F, Stephenne X, Sokal E, et al. Activity of the alpha-1 antitrypsin deficiency registry in Belgium. COPD 2015;12(Suppl. 1):10–4.

[12] Seersholm N, Dirksen A, Kok-Jensen A. Airways obstruction and two year survival in patients with severe alpha 1-antitrypsin deficiency. Eur Respir J 1994;7:1985–7.

[13] Gauvain C, Mornex JF, Pison C, Cuvelier A, Balduyck M, Pujazon MC, et al. Health-related quality of life in patients with alpha-1 antitrypsin deficiency: the French experience. COPD 2015;12(Suppl. 1):46–51.

[14] Bals R, Koczulla R, Kotke V, Andress J, Blackert K, Vogelmeier C. Identification of individuals with alpha-1-antitrypsin deficiency by a targeted screening program. Respir Med 2007;101:1708–14.

[15] Koczulla R, Bittkowski N, Andress J, Greulich T, Schroth S, Kotke V, et al. [The German registry of individuals with alpha-1-antitrypsin deficiency—a source for research on patient care]. Pneumologie 2008;62:655–8.

[16] Fähndrich S, Herr C, Greulich T, Seibert M, Lepper PM, Bernhard N, et al. Sex differences in alpha-1-antitrypsin deficiency lung disease-analysis from the German registry. COPD 2015;12(Suppl. 1):58–62.

[17] O'Brien ME, Pennycooke K, Carroll TP, Shum J, Fee LT, O'Connor C, et al. The impact of smoke exposure on the clinical phenotype of alpha-1 antitrypsin deficiency in Ireland: exploiting a national registry to understand a rare disease. COPD 2015;12(Suppl. 1):2–9.

[18] Ferrarotti I, Carroll TP, Ottaviani S, Fra AM, O'Brien G, Molloy K, et al. Identification and characterisation of eight novel SERPINA1 Null mutations. Orphanet J Rare Dis 2014;9:172.

[19] Balbi B, Fiasella F, Fabiano F. Alpha1-antitrypsin deficiency in a Z homozygote in Italian patient with bronchiectasis. Rassegna di Patologia dell'Apparato Respiratorio 1994;9:254–7.

[20] Luisetti M, Massi G, Massobrio M, Guarraci P, Menchicchi FM, Beccaria M, et al. A national program for detection of alpha 1-antitrypsin deficiency in Italy. Gruppo I.D.A. Respir Med 1999;93:169–72.

[21] Luisetti M, Ferrarotti I, Corda L, Ottaviani S, Gatta N, Tinelli C, et al. Italian registry of patients with alpha-1 antitrypsin deficiency: general data and quality of life evaluation. COPD 2015;12(Suppl. 1):52–7.

[22] Stolk J, Stockley RA, Stoel BC, Cooper BG, Piitulainen E, Seersholm N, et al. Randomised controlled trial for emphysema with a selective agonist of the γ-type retinoic acid receptor. Eur Respir J 2012;40:306–12.

[23] Fregonese L, Ferrari F, Fumagalli M, Luisetti M, Stolk J, Iadarola P. Long-term variability of desmosine/isodesmosine as biomarker in alpha-1-antritrypsin deficiency-related COPD. COPD 2011;8:329–33.

[24] Chorostowska-Wynimko J, Nizankowska-Mogilnicka E, Bakuła A, Górecka D, Kulus M, Grupa robocza Polskiego Towarzystwa Chórob Płuc. et al. [Diagnosis and treatment of patients with alpha-1 antitrypsin (alpha-1 AT) deficiency] [Article in Polish]. Pneumonol Alergol Pol 2010;78:348–55.

[25] Chorostowska-Wynimko J, Struniawski R, Sliwinski P, Wajda B, Czajkowska-Malinowska M. The national alpha-1 antitrypsin deficiency registry in Poland. COPD 2015;12(Suppl. 1):22–6.

[26] Vidal R, Miravitlles M, de Gracia X, Gallego B, Morell F. [Replacement therapy of emphysema caused by alpha 1-antitrypsin deficiency] [Article in Spanish]. Med Clin (Barc) 1991;96:180–2.

[27] Vidal R, Blanco I, Casas F, Jardí R, Miravitlles M. Committee on the National Registry of Individuals with Alpha-1 Antitrypsin Deficiency [Guidelines for the diagnosis and management of alpha-1 antitrypsin deficiency]. Arch Bronconeumol 2006;42(12):645–59.

[28] Casas F, Blanco I, Martínez MT, Bustamante A, Miravitlles M, Cadenas S, et al. Indications for active case searches and intravenous alpha-1 antitrypsin treatment for patients with alpha-1 antitrypsin deficiency chronic pulmonary obstructive disease: an update. Arch Bronconeumol 2015;51:185–92.

[29] Lara B, Miravitlles M. Spanish Registry of Patients with Alpha-1 Antitrypsin Deficiency: comparison of the characteristics of PISZ and PIZZ individuals. COPD 2015;12(Suppl. 1): 27–31.

[30] Laurell CB, Eriksson S. The serum alpha-l-antitrypsin in families with hypo-alpha-l-antitrypsinemia. Clin Chim Acta 1965;11:395–8.

[31] Laurell CB, Eriksson S. The electrophoretic α1-globulin pattern of serum in α1-antitrypsin deficiency. 1963. COPD 2013;10(Suppl. 1):3–8.

[32] Piitulainen E, Tanash HA. The clinical profile of subjects included in the Swedish National Register on individuals with severe alpha 1-antitrypsin deficiency. COPD 2015;12(Suppl. 1):36–41.

[33] Stockley RA. Antitrypsin Deficiency Assessment and Programme for Treatment (ADAPT): the United Kingdom Registry. COPD 2015;12(Suppl. 1):63–8.

[34] Bradi AC, Audisho N, Casey DK, Chapman KR. Alpha-1 antitrypsin deficiency in Canada: regional disparities in diagnosis and management. COPD 2015;12(Suppl. 1):15–21.

[35] The Alpha-1-Antitrypsin Deficiency Registry Study Group. Survival and FEV 1 decline in individuals with severe deficiency of alphal-antitrypsin. Amer J Respir Crit Care Med 1998;158:49–59.

[36] Stoller JK, Brantly M, Fleming LE, Bean JA, Walsh J. Formation and current results of a patient-organized registry for alpha(1)-antitrypsin deficiency. Chest 2000;118:843–8.

[37] Strange C, Monk R, Schwarz L, Walker D, Kumbhare S, Bieko T. The United States alpha-1 foundation research registry: genesis, impact and future. COPD 2015;12(Suppl. 1):42–5.

[38] Campos MA, Alazemi S, Zhang G, Wanner A, Sandhaus RA. Effects of a disease management program in individuals with alpha-1 antitrypsin deficiency. COPD 2009;6:31–40.

Chapter 15

Current Research and Future Perspectives

15.1 INTRODUCTION

After the discovery of the alpha-1 antitrypsin (AAT) deficiency (AATD) about of 50 years ago, there has been a significant progress in the knowledge of its pathophysiological mechanisms, as well as in clinical diagnosis and management of the diseases related to this rare inherited condition. A major advance for lung disease has been obtaining purified human plasma-derived AAT to treat deficient subjects with COPD, a therapy firstly approved in the United States and later extended to a growing number of developed countries [1]. However, intravenous augmentation therapy is a lifelong expensive, onerous, and time-consuming therapy. Therefore, although the potential efficacy of aerosol-delivered AAT on AATD-related lung inflammation has not been fully demonstrated yet [2], it is believed it could be a good option for an easier, cheaper, and more efficient delivery of AAT directly into the lungs [3].

On the other hand, in later years it has been demonstrated that AAT is a broad-spectrum protease inhibitor with antiinflammatory, antiapoptotic, and immunomodulatory properties, and there have been reported cases of systemic vasculitis, panniculitis, fibromyalgia, and asthma which responded to "of label" treatment with AAT infusions, in extreme situations where conventional therapy had previously failed [4–7]. All these facts, have suggested other potential applications of AAT in several diseases related and unrelated to the AATD [1]. In this context, some patents published in the last decade claim for the clinical use of AAT in pathological situations as diverse as autoimmune type-1 diabetes, graft vs host disease, graft protection, bacterial infections, radiation exposure, or fibromyalgia and chronic fatigue syndromes [8].

Additionally, in recent years some significant and bright advances in the achievement of alternative sources of AAT, other than human plasma, and in innovative strategies for AATD treatment have been reported, although most of them are still in development [9]. For example, it has been proven that wild-type donor hepatocytes can almost completely repopulate the liver of a Pi*Z mouse model of AATD [10]. If this "cellular" therapy could be applied to humans, its use could be an alternative approach to prevent or treat COPD and also liver disease.

In addition, recently the FDA approved several drugs that enhance autophagy, which could be valuable candidates for chemoprophylaxis of AATD liver disease [11].

Remarkably, AATD has been an attractive target for the development of gene therapy because (1) it is a single gene disorder; (2) the AAT coding sequence is relatively short; and (3) the AAT production from any cell or tissue capable of secreting the protein could be useful for augmentation therapy. Based on these considerations, attempts have been made to develop AAT therapies using nonviral gene transfer, gamma-retrovirus, recombinant adenovirus (rAd), and recombinant adeno-associated virus (rAAV) vectors [12]. In this context, other new therapeutic strategies, including gene silencing and the use of genetically engineered iPS-cell-derived hepatocytes, are currently under active investigation [13].

15.2 NEW STRATEGIES FOR AUGMENTATION THERAPY

15.2.1 Double Dose of Intravenous Alpha-1 Antitrypsin for Alpha-1 Antitrypsin Deficiency-Related COPD

With the approved dose of 60 mg/kg weekly, AAT intravenous augmentation therapy achieves a trough serum level of 11 μM, which is far below the level observed in non-AATD individuals. With the hypothesis that this serum level may be insufficient to protect the lung tissues from elastase, an ongoing randomized, placebo-controlled *S*tudy of *P*rol*A*stin-c *R*andomized *T*herapy with *A*lpha-1 augmentation (*acronym: SPARTA*; ClinicalTrials.gov Id: NCT01983241) has been designed. Its objective is to assess the lung tissue loss in adult patients with AATD-related pulmonary emphysema while receiving human AAT (Prolastin-C) versus placebo, with two separate doses of Prolastin-C (60 and 120 mg/kg) administered weekly over 3 years. The primary measure of efficacy [change from baseline whole-lung 15th percentile lung density (PD15)] is being determined by whole-lung computed tomography (CT) densitometry measured at total lung capacity. Secondary efficacy variables will be the evaluation of severe COPD exacerbations. Adverse events will be collected and documented [14].

Before SPARTA, a pilot trial assessed the safety and pharmacokinetic profile of weekly infusions of a 120 mg/kg dose of AAT in 30 adults with severe AATD, randomly assigned to weekly infusions of 60 or 120 mg/kg AAT (Prolastin-C) for 8 weeks before crossing over to the alternate dose for 8 weeks. The 120 mg/kg AAT weekly dose was considered to be safe and well tolerated, and provided more favorable physiologic AAT serum levels than the currently recommended 60 mg/kg dose [15].

15.2.2 Inhalation Therapy

Aerosolized AAT is an alternative therapeutic option to the intravenous route, and it is expected that with this approach local lung concentrations of AAT can be achieved at low doses, sparing both systemic delivery of AAT and cost [16]. Several classical studies in patients with cystic fibrosis have demonstrated the ability of aerosolized human AAT to inhibit neutrophil elastase and to decrease IL-8 concentration in the alveolar fluid, as well as to reduce neutrophil counts in sputum, with acceptable safety and tolerance [17–21].

Regarding AATD-associated COPD, there is currently underway a long-term randomized, double-blind, placebo-controlled trial, in 168 Pi*Z subjects treated with a high-purity liquid formulation (160 mg/day) of AAT nebulized from Kamada Lab., of which there are not available results yet (ClinicalTrials.gov Id: NCT01217671).

15.2.3 Recombinant/Transgenic Alpha-1 Antitrypsin

At present, only human plasma-derived AAT is available for clinical use. However, given the limited supply, the high cost, the potential infection risks, and the likely use of this product in a number of diseases other than the typical AATD-related COPD, there will be a need for new formulations of it in the future [22,23].

Recombinant AAT has been derived from plants, yeast, fungi, insect cells, bacteria and mammalian cells, and modified and conjugated with polyethylene glycol to prolong its half-life in blood [24–30], but none of these formulations has managed to match the traditional plasma-derived affinity-purified form of human AAT. On the other hand, the not yet resolved difficulty in achieving an adequate protein glycosylation can shorten its blood half-life, and may cause functional impairment and adverse immune-mediated responses [31,32].

By use of transgenic sheep that carry the genetic sequence for human AAT under the control of mammary gland promoters, a biologically active human AAT was generated in the sheep milk. By using this method, it was estimated that a population of 4500 sheep would be able to provide 5000 kg of human AAT annually [33–35]. This protein was later purified and introduced to humans by intravenous (i.v.) infusions. Unfortunately, a clinical trial in humans produced intolerable allergic reactions (i.e., i.v. administration caused febrile allergic reactions, and aerosolized, episodes of "hypersensitivity pneumonitis" for antibodies to residual traces of ovine alpha-1 antichymotrypsin) that forced its discontinuation [36].

More recently, a recombinant AAT formulation, derived from healthy immortalized human embryonic retinal cells (PER.C6 cells), has been developed, and it has achieved a near identical glycosylation when compared to the M plasma-derived AAT. This recombinant AAT has also similar pharmacokinetics and antielastase activity as the native AAT in vitro. Therefore, although further studies are needed to demonstrate the clinical safety and efficacy of this (and other similar products currently in development) recombinant human AAT, it might offer a reliable source for augmentation therapy [37].

15.3 STRATEGIES FOR INCREASING ENDOGENOUS PRODUCTION OF ALPHA-1 ANTITRYPSIN

15.3.1 Chemical Chaperones

Chemical chaperones can facilitate folding of multiple misfolded proteins nonselectively, and thus they have been investigated as potential therapeutic drugs for AATD treatment [38]. Some of them have been effective in cell and animal models of disease. Specifically, in a model cell culture system, 4-phenylbutyric acid (PBA) mediated a marked increase in secretion of functionally active Z-AAT. Moreover, this drug also mediated an increase in blood levels of human Z-AAT when administered orally to a transgenic mouse model of Z-AAT, with blood levels reaching 20%–50% of the levels present in normal humans [39]. In view of these positive findings, the authors concluded that because PBA had been used safely in children with urea-cycle disorders for years, this drug could be considered as a candidate for chemoprophylaxis of both liver and lung disease in AATD. However, a preliminary open label study evaluating the effect of 14 days of oral 4-PBA therapy on AAT blood levels, in 10 patients with AATD, has been disappointing, founding no significant increase in AAT blood level associated with 4-PBA administration, and in contrast significant symptomatic and metabolic side effects [40]. Nevertheless, it may be interesting to further investigate PBA if more favorable formulations become available.

15.3.2 Suberoylanilide Hydroxamic Acid

More recently, suberoylanilide hydroxamic acid (SAHA) has shown to be able to increase secretion of Z-AAT in cell line models of AATD, and this effect was proposed to be mediated by inhibition of the histone deacetylase HDAC7 [41]. However, this drug was not tested in vivo, and it was not entirely clear whether the increase in Z-AAT in extracellular fluid could be mostly attributable to an increase in synthesis. Even if the results were consistent with increases in protein synthesis as well as an increase in translocation through the secretory pathway, the effect of this drug would be associated with increased accumulation of Z-AAT in the ER and therein possibly with increased proteotoxicity [11].

15.4 LUNG CELLULAR REGENERATION WITH *TRANS*-RETINOIC ACID

In animal models of emphysema, palovarotene (a selective agonist of the retinoic acid receptor gamma) reduced inflammation and repaired the structural damage of damaged lungs. However, in humans, a randomized, double-blind, placebo-controlled study (*acronym: REPAIR, Retinoid treatment of Emphysema in Patients on the α(1)-antitrypsin International Registry*) in 129 Pi*Z individuals treated with 5 mg orally/day/1 year versus 133 Pi*Z treated with placebo, failed to show significant benefit on lung density in moderate-to-severe emphysema secondary to AATD [42].

15.5 STRATEGIES TO BLOCK INTRACELLULAR POLYMERIZATION

15.5.1 Small Molecules

Polymerization is the central feature of the AATD. Sequential insertion of the loop of one mutant AAT molecule into β-sheet A of a neighbor, to form a loop-sheet dimer, and then longer species linking more molecules, explains the formation of elongated polymers [43,44]. This thorough understanding is being used to develop novel therapeutic strategies with the ultimate goal of developing a "cure" for AATD [45]. Indeed, polymerization can be blocked by peptides that mimic the AAT reactive loop sequence and so compete for binding to the insertion site in the β-sheet A [46–49]. However, the challenge is to deliver these peptides to the human liver cells [44].

In addition, the hydrophobic pocket of the AAT surface bounded by strand 2A and helices D and E [50] has been used as a target for small molecules directed to block polymer formation. Some of these compounds were able to reduce or completely block polymer formation in vitro and in cell models of disease [51], although these short peptides have not been tested in animal models yet [52].

15.5.2 Intrabodies to Block Intracellular Polymerization and Increase Secretion

More recently the David Lomas team, using advanced monoclonal antibody technology, has identified some intrabodies (i.e., antibodies that work within the cell to bind to an intracellular protein) that can block polymer formation in vitro at a 1:1 molar ratio [53]. This novel approach could reveal new target sites for small molecules intervention, able to block the transition to aberrant polymers without compromising the limited inhibitory activity of the Z-AAT [54].

15.6 GENE THERAPY

Gene therapy is the therapeutic delivery of nucleic acids into the patient's cells to treat disease. Gene therapy strategies for AATD have mostly focused on gene augmentation as a means to treat AAT lung disease. The classical approach of inserting normal genes into cells of patients with the AAT single-gene mutation has been investigated in the hope of achieving therapeutic AAT levels to preserve lung integrity. However, gene therapies utilizing this approach will not prevent the production of mutant Z-AAT, and they will not avoid the protein accumulation into hepatocytes.

TABLE 15.1 Clinical Trials of Gene Therapy in AATD

Phase	Vector	Route	Number of patients	Main results/ comments	Year, author (references)
I	Plasmid–cationic liposome complex	Intranasal	5	AAT concentration increased in nasal lavage to 1/3 normal, but returned to baseline at day 14	2000, Brigham [55]
I	rAAV2-AAT	IM	12	Only one subject had low serum AAT expression 30 days after an injection of 2.1×10^{13} vector genomes. Anti-AAV2 capsid antibodies rose after vector injection	2006, Brantly [56]
I	rAAV1-AAT	IM	9	AAT in subtherapeutic levels (0.1% of normal) in two subjects after 1 year. All patients developed anti-AAV2 capsid antibodies by day 14.	2009, Brantly [57]
II	rAAV1-AAT (herpes simplex virus method)	IM	9	Dose-dependent increase of AAT levels, but still subtherapeutic. Transient elevation of creatine kinase	2011, Muller and Flotte [58]

AAT, Alpha-1 antitrypsin; AATD, alpha-1 antitrypsin deficiency; IM, intramuscular; rAAV, recombinant adeno-associated virus.

Gene-based therapy for AATD has been developed using multiple different vector systems, including nonviral gene transfers (e.g., cationic liposomes), gamma-retrovirus, rAd, and several rAAV serotypes. These vectors have been administered by different routes of delivery, but in practice, no gene therapies are currently licensed for use in AATD.

To date, there have been three published phase I clinical trials (one of cationic liposome, one of rAAV2, and one of rAAV1) as well as one phase II trial (with rAAV1) [55–58] (Table 15.1). The results of the latter trial, while promising, demonstrated levels that were only 3%–5% of the target range. This indicates the need to further increase the dose of the vector. The remaining studies have also failed to provide satisfactory results due to the low serum levels of AAT achieved, the short-time expression of AAT, and the frequent development of antibodies against vector viruses [12].

15.7 OLIGONUCLEOTIDE-BASED THERAPEUTICS TARGETING THE AAT GENE

Oligonucleotide-based therapeutics are an emerging class of small drugs completely synthetic in origin, which do not integrate into the host genome, and have a defined duration of therapeutic activity after which effects recover to baseline. In addition, they offer a high degree of specificity for targeting any gene in the genome reducing off-target effects. Oligonucleotide-based therapeutics include: short (or small) interfering RNA (siRNA) that degrade a target messenger RNA (mRNA) through RNA-induced silencing complex (RISC) mediated RNA interference (RNAi); anti-microRNAs (anti-miRs) that target microRNAs (miRNAs); miRNA mimics, that regulate target mRNA; antisense oligonucleotides (ASOs) that work through ribonuclease H (RNase H) mediated mRNA decay; mRNA upregulation, by targeting long noncoding RNAs; and oligonucleotides-induced alternative splicing.

While there have been many reports of gene knockdown and gene modulation in cell lines and mice showing promising results in preclinical models with all these methods, very few have advanced to clinical stages [59,60]. The main obstacle to date has been the safe and effective intracellular delivery of these compounds in humans because their action requires direct intranuclear interaction with DNA/RNA within the target cell, a very difficult to overcome barrier [61].

An ASO is a synthesized short strand of deoxyribonucleotide analogue, which hybridizes with the complementary mRNA in a sequence-specific manner. The ASO-mRNA heteroduplex triggers RNase H activity, leading to mRNA degradation, induces translational arrest of ribosomal activity, interferes with mRNA maturation or destabilizes pre-mRNA in the nucleus, resulting in downregulation of a target protein expression [62]. ASOs targeting human AAT gene efficiently reduce levels of both short and long human AAT transcripts in vitro and in transgenic mice, providing a potentially novel therapy for AATD liver disease [63]. Specifically, a recent study has shown that Pi*Z transgenic mouse strain expresses a human AAT transgene that contains the AATD-associated Pi*Z mutation, and exhibit many AATD-liver disease characteristics, including AAT protein aggregates, increased hepatocyte death, and liver fibrosis. AAT-ASO administration in these animals stopped liver disease progression after short-term treatment, reversed liver disease after long-term treatment, and prevented liver disease in young animals. Additionally, ASO treatment markedly decreased liver fibrosis in this mouse model. Moreover, administration of AAT-ASO in nonhuman primates led to an approximately 80% reduction in levels of circulating normal AAT, demonstrating potential for this approach in humans with AATD [64].

Gene knockdown by RNA interference (RNAi) represents a promising therapy for AATD-related liver disease [65]. Two pharmaceutical companies (Arrowhead Research Corporation and Alnylam Pharmaceuticals Inc.) have developed RNAi-based approaches to silence Z-AAT production in hepatocytes, using small interfering (siRNA) constructs targeted against mRNA encoding human AAT. A single dose escalation study in healthy volunteers and patients has been designed to knockdown the hepatic production of the mutant Z-AAT protein, and is now in a phase I/II clinical trial (NCT02503683).

For diseases that are attributable to both loss-of-function and gain-of-toxicity, as is the case of AATD, a dual vector should be designed to simultaneously knockdown the toxic product from the mutated gene and to replace it with a normal gene. Therefore, vectors capable of allele-specific down-regulation of Z-AAT with simultaneous augmentation of M-AAT have demonstrated to be efficacious, stable, and safe in a mouse model, but this is still in initial phases of development in humans [66]. In this context, several researches have reported the manufacture of MiRNAs that target de AAT gene [67]. Using the same rAAV vector they incorporated miRNA sequences that target de Pi*Z gene while also driving the expression of a wild-type Pi*M gene (which is resistant to the MiRNA-mediated knockdown), getting the desired effect of reducing circulating Z-AAT while increasing M-AAT protein in transgenic mice [68].

15.8 CORRECTION OF THE Z-AAT GENE IN AUTOLOGOUS INDUCED PLURIPOTENT STEM CELLS

Induced pluripotent stem cells (also known as iPS cells or iPSCs) are a type of pluripotent stem cells that can be generated directly from adult cells genetically reprogrammed to an embryonic-like state, by transfer of nuclear contents into oocytes or by fusion with embryonic stem cells [69–71].

The creation of iPSCs from adult human cells was a major scientific achievement 10 years ago, with potential applications in the fields of regenerative medicine (because they offer the prospect of generating unlimited quantities of cells for autologous transplantation), disease modeling and drug discovery, with potential application in treatments for a broad range of disorders [72]. Specifically, these cells provide an invaluable resource for modeling AATD and other genetic diseases [73,74].

Remarkably, the successful genetic correction of human iPSCs derived from the skin fibroblasts from a Pi*Z homozygous patient in vitro has been reported. In these dermal-derived iPSCs, the Pi*Z mutant base was removed and replaced by a normal Pi*M sequence (by transposon excision and cut-and-paste procedures) to generate a normal Pi*M gene. These corrected iPSCs could then be cloned and differentiated into hepatocytes in vitro, and could eventually be transplanted into the liver of the original person, where the corrected

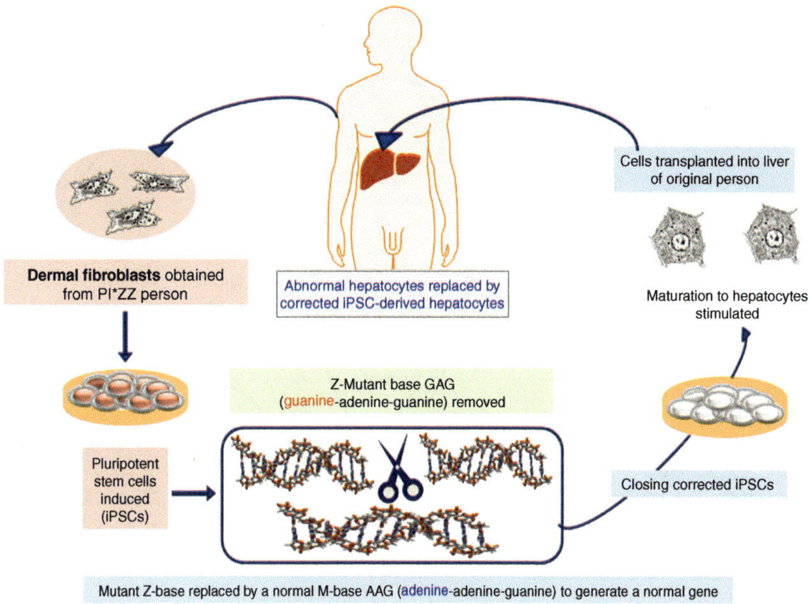

FIGURE 15.1 Person with alpha-1 antitrypsin (AAT) deficiency treated with autologous corrected induced pluripotent stem cells (iPSCs). This therapy consists of correction of the mutated gene Z by genetic engineering, combining zinc finger nuclease and transposons "piggy back" techniques. Schematically, the process is as follows: (1) skin fibroblasts of the person with AATD are taken. (2) These fibroblasts are induced to become iPSCs. (3) In these iPSCs, the mutated nucleotide DNA base Z gene is deleted, and the correct nucleotide M is introduced to generate a normal gene M. (4) The corrected iPSCs are cloned and differentiated into hepatocytes. (5) These iPSC-derived hepatocytes are transplanted from the liver of the same person. (6) The hepatocytes derived from the corrected iPSCs replace abnormal hepatocytes, virtually eliminating the chance of rejection.

iPSC-derived hepatocytes might replace the abnormal hepatocytes (Fig. 15.1). This approach has been demonstrated to properly work in vitro and in vivo in mice, as well as in human cells from individuals with hereditary AATD, providing a sustained serum expression of M-AAT [75].

However, although this strategy has the potential to correct both the loss-of-function and gain-of-function mechanisms, and would have the advantage of obviating the need for immunosuppression, the risk of developing alterations of the nucleotide sequence of the genome (including cancer mutations) during iPSC culture, as well as the ethical concerns still unanswered, has precluded its clinical use to date [13].

15.9 AUTOPHAGY-ENHANCING DRUGS

The ubiquitin–proteasome and autophagy–lysosomal pathways are the two major routes for protein and organelle clearance in eukaryotic cells. Proteasomes predominantly degrade short-lived nuclear and cytosolic proteins.

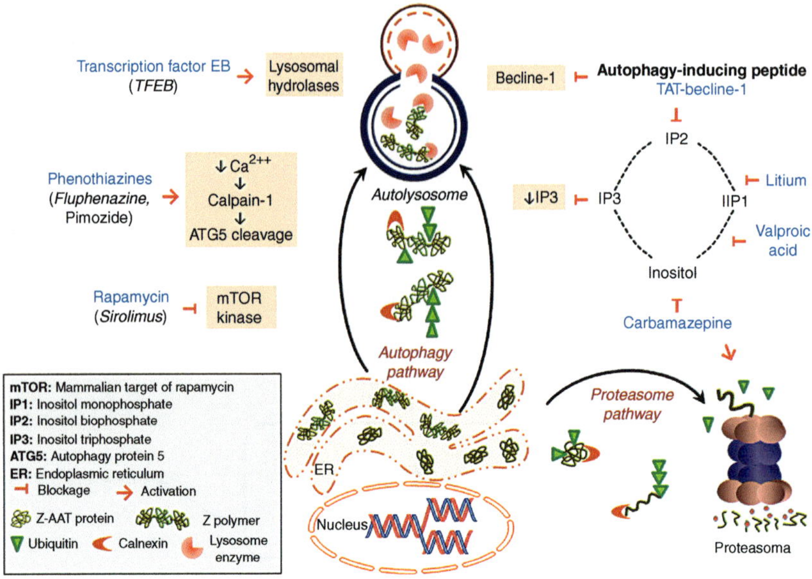

FIGURE 15.2 Degradative pathways of the Z alpha-1 antitrypsin protein and Z polymers, and mechanismo of action of proautophagic drugs.

The bulk degradation of cytoplasmic proteins or organelles is largely mediated by autophagy. It involves the formation of double-membrane structures, called autophagosomes, which fuse with lysosomes to form autolysosomes. Autophagy substrates generally have long half-lives [76]. In subjects with AATD, the proteasomal pathway is capable of degrading soluble forms of Z-AAT molecules, but at higher levels of expression, as Z-AAT accumulates in polymers and insoluble aggregates, autophagy becomes critical for polymers degradation.

Currently, the only suitable pharmacologic strategy for upregulating autophagy in mammalian cells is to use rapamycin (Sirolimus), an immunosuppressant used to prevent organ transplantation rejection. Rapamycin inhibits the mammalian target of rapamycin (mTOR), a serine/threonine protein kinase negative regulator of autophagy (Fig. 15.2). Autóphagy activity can be enhanced by rapamycin when given in weekly dose pulses, rather than as a daily regimen. In a mouse model of Z-AATD, weekly doses of rapamycin increased the number of autophagic vacuoles and reduced the accumulation of intrahepatic polymerized Z-AAT and liver injury; additionally, it also reduced other markers of hepatocellular injury, including cleavage of caspase 12, and hepatic fibrosis on Pi*Z mice in vivo [77].

Inositol depletion is a common mechanism for mood stabilizing drugs like lithium, carbamazepine and valproic acid [78,79]. For example, lithium (a mood-stabilizing drug for bipolar disorder) induces autophagy by inhibiting inositol monophosphatase, leading to free inositol depletion, and this in turn decreasing inositol-1 triphosphate (IP_3) levels [80]. Carbamazepine is an anticonvulsant and mood stabilizer drug, extensively used in clinical practice, which enhances autophagy via inositol monophosphatase inhibition, and additively also raises the clearance of mutant Z-AAT molecules by intensifying the proteasoma activity. Administered orally to an AATD-associated liver disease Pi*Z mice, the drug decreased the hepatic load of Z-AAT and hepatic fibrosis [81]. Since carbamazepine is approved by the FDA for use in clinical practice, it has been moved into a Phase II/III ongoing trial for patients with severe liver disease related to AATD (NCT01379469).

The autophagy regulator transcription factor EB (TFEB) is a protein which coordinates expression of lysosomal hydrolases, membrane proteins and genes involved in autophagy [82]. In cultured cells, TFEB induced lysosomal biogenesis and increased the degradation of complex molecules, such as glycosaminoglycans and the pathogenic protein that causes Huntington disease. Therefore, TFEB gene transfer may be a novel strategy for treatment of liver disease in SERPINA1 deficiency, and may have broad applications for therapy of human disorders due to intracellular accumulation of toxic proteins. In the liver of a transgenic mouse model of SERPINA1 deficiency, liver-directed gene transfer of TFEB resulted in marked reduction of SERPINA1 polymers, apoptosis and fibrosis [83,84].

Furthermore, two antipsychotic drugs, from the phenothiazine family, used in schizophrenia and chronic psychosis, fluphenazine and pimozide, have also been shown to enhance autophagy degradation of the protein huntingtin (the protein mutated in Huntington's disease) [85]. Although currently not well clarified, it appears that phenothiazines may enhance autophagic degradation reducing the intracellular calcium level, which leads to diminished calpain-1-mediated cleavage of the autophagy protein 5 (ATG5), an ubiquitin ligase necessary for autophagy, due to its role in autophagosome elongation [11].

Finally, a novel autophagy-inducing 8-amino acid peptide, termed Tat-beclin 1 enhances autophagy by directly interacting with the autophagy regulatory factor beclin 1, a protein which interacts with either B-cell lymphoma 2 (BCL-2) regulator proteins of apoptosis or phosphatidylinositol-3-kinases (PI3k), playing a critical role in the regulation of both autophagy and cell death [86]. In laboratory experiments, Tat–beclin 1 decreased the accumulation of polyglutamine protein aggregates and the replication of several pathogens (including HIV-1) in vitro, and reduced mortality in mice infected with chikungunya or West Nile virus. These findings suggest that Tat-beclin 1 could potentially be used as a therapeutic agent in AATD, and also in other diseases caused by aggregation of proteins and possibly for some infectious diseases [86]. These findings further validate the concept that enhancing autophagy is a plausible therapeutic approach for ATD-associated liver disease, although again the challenge will be to demonstrate efficacy in man [11].

15.10 LIVER TRANSPLANTATION

Since transplanted hepatocytes have a selective proliferative advantage over endogenous hepatocytes and can eventually supplant the latter in a diseased liver, cell transplantation therapy for AATD-related liver cirrhosis has also been considered. This type of "cellular" therapy could theoretically prevent or treat liver disease and prevent COPD at the same time [11].

Of note, in a transgenic mice expressing human Z-AAT, wild-type donor hepatocytes replaced 20%–98% of mutant host hepatocytes, and repopulation was accelerated by injection of an adenovector expressing hepatocyte growth factor. Spontaneous hepatic repopulation with engrafted hepatocytes occurred in the Z-AAT-expressing mice even in the absence of severe liver injury. Donor cells replaced both globule-containing and globule-devoid cells, indicating that both types of host hepatocytes display impaired proliferation relative to wild-type hepatocytes. These results suggest that wild-type hepatocyte transplantation may be therapeutic for AAT-Z liver disease and may provide an alternative to protein replacement for treating emphysema in AAT-Pi*Z individuals [10].

15.11 NEW THERAPEUTIC INDICATIONS OF ALPHA-1 ANTITRYPSIN

In recent years, an impressive variety of in vitro and in vivo experimental studies, and some small case series have found that AAT is a multifunctional antiprotease, antinflammatory, immunomodulatory, antiinfective, and tissue-repair molecule, with potential biological properties for treating many diseases, both inside and outside the field of AATD [4,22,23].

Despite the fact that its real usefulness in clinical practice has yet to be demonstrated, in recent years they have been published in the United States an increasing number of patents claiming the use of the AAT (as monotherapy or associated to other standardized drugs) for pathological situations as diverse as fibromyalgia and chronic fatigue syndrome, diabetes mellitus, graft vs host disease (GvHD), allograft and xenograft transplantation, bacterial infections, or radiation exposure [87–96], (Table 15.2).

In addition, there are currently 10 clinical trials registered in the U.S. ClinicalTrials.gov registry, among which 3 are ongoing trials with weekly i.v. infusions of AAT in young patients with type 1 diabetes. The remaining registered clinical trials are in various stages of development, and cover several pathologies, including GvHD, neuromyelitis optica, cardiac surgery, and organ transplantation of kidneys, pancreatic islets, and lungs (Table 15.3).

TABLE 15.2 Patents (2007–2016) on Novel Diseases to be Treated by AAT, Listed by Chronological Order of Publication

Inventor(s)	Purpose	Year of publication (references)
Blanco	Use of AAT for the treatment of fibromyalgia	2007 [87]
Lewis et al.	Use of AAT for the promotion of graft survival and the treatment and prevention of GvHD	2011 [88]
Lewis et al.	Use of AAT for the treatment of type 1 diabetes	2012 [89]
Shapiro	Use of AAT for the treatment diabetes	2012 [90]
García-Quintana and Alegre-Martín	Use of AAT for the treatment of cognitive impairment associated with chronic fatigue syndrome	2012 [91]
Lewis et al.	Use of AAT and temporal T-cell depletion for the prevention of allograft and xenograft rejection	2013 [92]
Shapiro	Use of AAT Fc fusion molecule for the treatment and prevention of bacterial and mycobacterial infections, including *B. anthracis*	2014 [93]
Shapiro et al.	Use of AAT for the treatment of GvHD	2014 [94]
Shaphiro et al.	Compositions and methods related to graft versus host disease and treatments thereof	2014 [95]
Crapo et al.	Use of AAT to ameliorate the adverse effects of radiation exposure in patients	2015 [96]

Abbreviations: AAT, Alpha-1 antitrypsin; GvHD, graft versus host disease.

TABLE 15.3 Active Clinical Trials With AAT for "Non-Alpha Patients" Listed by Chronological Order

ClinicalTrials.gov	Title and current state of development	✓ Sponsor; • collaborator/s	First received/ last verified
NCT01319331	• The Effects of Alpha-1 Antitrypsin (AAT) on the Progression of Type 1 Diabetes • *Ongoing, but not recruiting participants.*	✓ University of Colorado, Denver. • *Omni Bio Pharmaceutical, Inc.*	March 2011/ October 2015
NCT01523821	• Alpha-1 antitrypsin in treating patients with acute graft versus host disease (GvHD) • *Recruiting participants.*	✓ Fred Hutchinson Cancer Research Center. • *National Cancer Institute (NCI) and* • *National Heart, Lung, and Blood Institute (NHLBI)*	January 2012/ May 2016
NCT01700036	• A pilot study of AAT in steroid refractory acute GvHD. • *Recruiting participants.*	✓ University of Michigan Cancer Center • *CSL Behring.* • *The Leukemia and Lymphoma Society.*	October 2012/ June 2016
NCT02005848	• Phase II Study to Evaluate the Efficacy and Safety of Glassia® in Type-1 Diabetes. • *Ongoing, but not recruiting participants.*	✓ Kamada, Ltd.	November 2013/ April 2016
NCT02087813	• Pilot Study of alpha 1-antitrypsin to Treat Neuromyelitis Optica Relapses (A1AT for NMO) • *The recruitment status of this study is unknown because the information has not been verified recently.*	✓ Stanford University	March 2014/ March 2014
NCT02093221	• Study of Human Plasma-Derived Alpha1-Proteinase Inhibitor in Subjects With New-Onset Type 1 Diabetes Mellitus. • *Ongoing, but not recruiting participants*	✓ Grifols Therapeutics Inc.	March 2014/ February 2016
NCT02191839	• Single Dose Administration of Alpha-1 Anti-Trypsin for the Amelioration of Organ Injury in Patients Undergoing Cardiac Surgery. • *Not yet open for participant recruitment.*	✓ Soroka University Medical Center	July 2014/ July 2014
NCT02464878	• Multicenter Trial of the Effect of AAT on Islet Transplant Engraftment and Durability after Renal Transplant. • *Not yet open for participant recruitment.*	✓ Massachusetts General Hospital. • *National Institute of Diabetes and Digestive and Kidney Diseases (NIDDK)* • *University of Iowa*	June 2015/March 2016
NCT02520076	• Aralast NP in islet transplant • *Currently recruiting participants.*	✓ University of Alberta. • *Baxalta US Inc.*	August 2015/ March 2016
NCT02614872	• Study to Evaluate the Safety and Efficacy of Intravenous Glassia® Treatment in Lung Transplantation. • *Recruiting participants.*	✓ Kamada, Ltd.	November 2015/ June 2016

REFERENCES

[1] de Serres F, Blanco I. Role of alpha-1 antitrypsin in human health and disease. J Intern Med 2014;276:311–35.

[2] Franciosi AN, McCarthy C, McElvaney NG. The efficacy and safety of inhaled human α-1 antitrypsin in people with α-1 antitrypsin deficiency-related emphysema. Expert Rev Respir Med 2015;9:143–51.

[3] Häussermann S, Winnips C, Edelman J, Kappeler D, Herpich C, Ehlich H, et al. Lung deposition of alpha1-proteinase inhibitor (human) (A1-PI[H]) inhalation solution using two inhalation modes of the I-neb adaptive aerosol delivery (AAD) system in healthy subjects and subjects with cystic fibrosis. J Aerosol Med Pulm Drug Deliv 2016;29:242–50.

[4] Blanco I, Lara B, de Serres F. Efficacy of alpha1-antitrypsin augmentation therapy in conditions other than pulmonary emphysema. Orphanet J Rare Dis 2011;6:14.

[5] Blanco I, Lipsker D, Lara B, Janciauskiene S. Neutrophilic panniculitis associated with alpha-1-antitrypsin deficiency: an update. Br J Dermatol 2016;174:753–62.

[6] Blanco I, Canto H, Flóres J, Camblor C, Cárcaba V, de Serres FJ, et al. Long-term augmentation therapy with alpha-1 antitrypsin in an MZ-AAT severe persistent asthma. Monaldi Arch Chest Dis 2008;69:178–82.

[7] Blanco I, Canto H, de Serres FJ, Fernandez-Bustillo E, Rodríguez MC. Alpha1-antitrypsin replacement therapy controls fibromyalgia symptoms in 2 patients with PI ZZ alpha1-antitrypsin deficiency. J Rheumatol 2004;31:2082–5.

[8] Lior Y, Geyra A, Lewis EC. Therapeutic compositions and uses of alpha1-antitrypsin: a patent review (2012–2015). Expert Opin Ther Pat 2016;26:581–9.

[9] Lomas DA, Hurst JR, Gooptu B. Update on alpha-1 antitrypsin deficiency: new therapies. J Hepatol 2016;65:413–24.

[10] Ding J, Yannam GR, Roy-Chowdhury N, Hidvegi T, Basma H, Rennard SI, et al. Spontaneous hepatic repopulation in transgenic mice expressing mutant human α1-antitrypsin by wild-type donor hepatocytes. J Clin Invest 2011;121:1930–4.

[11] Ghouse R, Chu A, Wang Y, Perlmutter DH. Mysteries of α1-antitrypsin deficiency: emerging therapeutic strategies for a challenging disease. Dis Model Mech 2014;7:411–9.

[12] Mueller C, Flotte TR. Gene-based therapy for alpha-1 antitrypsin deficiency. COPD 2013;10(Suppl. 1):44–9.

[13] Sandhaus RA. Gene therapy meets stem cells. N Engl J Med 2012;366:567–9.

[14] Sorrells S, Camprubi S, Griffin R, Chen J, Ayguasanosa J. SPARTA clinical trial design: exploring the efficacy and safety of two dose regimens of alpha1-proteinase inhibitor augmentation therapy in alpha1-antitrypsin deficiency. Respir Med 2015;109:490–9.

[15] Campos MA, Kueppers F, Stocks JM, Strange C, Chen J, Griffin R, et al. Safety and pharmacokinetics of 120 mg/kg versus 60 mg/kg weekly intravenous infusions of alpha-1 proteinase inhibitor in alpha-1 antitrypsin deficiency: a multicenter, randomized, double-blind, crossover study (SPARK). COPD 2013;10:687–95.

[16] Monk R, Graves M, Williams P, Strange C. Inhaled alpha 1-antitrypsin: gauging patient interest in a new treatment. COPD 2013;10:411–5.

[17] Hansen G, Schuster A, Zubrod C, Wahn V. Alpha 1-proteinase inhibitor abrogates proteolytic and secretagogue activity of cystic fibrosis sputum. Respiration 1995;62:117–24.

[18] Tebbutt SJ. Technology evaluation: transgenic alpha-1-antitrypsin (AAT), PPL therapeutics. Curr Opin Mol Ther 2000;2:199–204.

[19] Martin SL, Downey D, Bilton D, Keogan MT, Edgar J, Elborn JS. Safety and efficacy of recombinant alpha1-antitrypsin therapy in cystic fibrosis. Ped Pulmonol 2006;41:177–83.

[20] Griese M, Latzin P, Kappler M, Weckerle K, Heinzlmaier T, Bernhardt T, et al. alpha1-Antitrypsin inhalation reduces airway inflammation in cystic fibrosis patients. Eur Respir J 2007;29:240–50.

[21] Kerem E, Bauer S, Strauss P, Jaffe N, Armoni S, Pugatsch T, et al. Safety/efficacy of inhaled human alpha-1 antitrypsin (AAT) in CF: a phase II clinical study. J Cyst Fibr 2009;8:S25.

[22] Lewis EC. Expanding the clinical indications for α(1)-antitrypsin therapy. Mol Med 2012;18:957–70.

[23] Wanner A, Arce AD, Pardee E. Novel therapeutic uses of alpha-1 antitrypsin: a window to the future. COPD 2012;9:583–8.

[24] Casolaro MA, Fells G, Wewers M, Pierce JE, Ogushi F, Hubbard R, et al. Augmentation of lung antineutrophil elastase capacity with recombinant human alpha-1-antitrypsin. J Appl Physiol 1987;63:2015–23.

[25] Archibald AL, McClenaghan M, Hornsey V, Simons JP, Clark AJ. High-level expression of active recombinant human α1-antitrypsin in the milk of transgenic mice. Proc Natl Acad Sci USA 1990;87:5178–82.

[26] Massoud M, Bischoff R, Dalemans W, Pointu H, Attal J, Schultz H, et al. Expression of active recombinant human alpha 1-antitrypsin in transgenic rabbits. J Biotechnol 1991;18:193–203.

[27] Wright G, Carver A, Cottom D, Reeves D, Scott A, Simons P, et al. High level expression of active human alpha-1-antitrypsin in the milk of transgenic sheep. Biotechnology 1991;9: 830–4.

[28] Niemann H, Kues WA. Application of transgenesis in livestock for agriculture and biomedicine. Anim Reprod Sci 2003;79:291–317.

[29] Karnaukhova E, Ophir Y, Golding B. Recombinant human alpha-1 proteinase inhibitor: towards therapeutic use. Amino Acids 2006;30:317–32.

[30] Cantin AM, Woods DE, Cloutier D, Dufour EK, Leduc R. Polyethylene glycol conjugation at Cys232 prolongs the half-life of alpha1 proteinase inhibitor. Am J Respir Cell Mol Biol 2002;27:659–65.

[31] Yu SD, Gan JC. Effects of progressive desialylation on the survival of human plasma alpha1-anti-trypsin in rat circulation. Int J Biochem 1978;9:107–15.

[32] Busby TF, Yu SD, Gan JC. Radioactive labeling of alpha1-antitrypsin, trypsin, and chymotrypsin by reductive methylation: properties of the labeled derivatives. Arch Biochem Biophys 1977;184:267–75.

[33] Carver A, Wright G, Cottom D, Cooper J, Dalrymple M, Temperley S, et al. Expression of human alpha 1 antitrypsin in transgenic sheep. Cytotechnology 1992;9:77–84.

[34] Carver AS, Dalrymple MA, Wright G, Cottom DS, Reeves DB, Gibson YH, et al. Transgenic livestock as bioreactors: stable expression of human alpha-1-antitrypsin by a flock of sheep. Biotechnology 1993;11:1263–70.

[35] Ziomek CA. Commercialization of proteins produced in the mammary gland. Theriogenology 1998;49:139–44.

[36] Spencer LT, Humphries JE, Brantly ML. Transgenic Human Alpha 1-Antitrypsin Study Group. Antibody response to aerosolized transgenic human alpha1-antitrypsin. N Engl J Med 2005;352:2030–1.

[37] Ross D, Brown T, Harper R, Pamarthi M, Nixon J, Bromirski J, et al. Production and characterization of a novel human recombinant alpha-1-antitrypsin in PER.C6 cells. J Biotechnol 2012;162:262–73.

[38] Devlin GL, Parfrey H, Tew DJ, Lomas DA, Bottomley SP. Prevention of polymerization of M and Z a1-antitrypsin (a1-AT) with Trimethylamine NOxide. Implications for the treatment of a1-AT deficiency. Am J Respir Cell Mol Biol 2001;24:727–32.

[39] Burrows JA, Willis LK, Perlmutter DH. Chemical chaperones mediate increased secretion of mutant alpha 1-antitrypsin (alpha 1-AT) Z: A potential pharmacological strategy for prevention of liver injury and emphysema in alpha 1-AT deficiency. Proc Natl Acad Sci USA 2000;97:1796–801.

[40] Teckman JH. Lack of effect of oral 4-phenylbutyrate on serum alpha-1-antitrypsin in patients with alpha-1-antitrypsin deficiency: a preliminary study. J Pediatr Gastroenterol Nutr 2004;39:34–7.

[41] Bouchecareilh M, Hutt DM, Szajner P, Flotte TR, Balch WE. Histone deacetylase inhibitor (HDACi) suberoylanilide hydroxamic acid (SAHA)-mediated correction of α1-antitrypsin deficiency. J Biol Chem 2012;287:38265–78.

[42] Stolk J, Stockley RA, Stoel BC, Cooper BG, Piitulainen E, Seersholm N, et al. Randomised controlled trial for emphysema with a selective agonist of the γ-type retinoic acid receptor. Eur Respir J 2012;40:306–12.

[43] Lomas DA, Evans DL, Finch JT, Carrell RW. The mechanism of Z a1-antitrypsin accumulation in the liver. Nature 1992;357:605–7.

[44] Lomas DA, Hurst JR, Gooptu B. Update on alpha-1 antitrypsin deficiency: new therapies. J Hepatol 2016;65:413–24.

[45] Lomas DA. Twenty years of polymers: a personal perspective on alpha-1 antitrypsin deficiency. COPD 2013;10(Suppl. 1):17–25.

[46] Mahadeva R, Dafforn TR, Carrell RW, Lomas DA. Six-mer peptide selectively anneals to a pathogenic serpin conformation and blocks polymerisation: implications for the prevention of Z a1-antitrypsin related cirrhosis. J Biol Chem 2002;277:6771–4.

[47] Parfrey H, Dafforn TR, Belorgey D, Lomas DA, Mahadeva R. Inhibiting polymerisation: new therapeutic strategies for Z a1-antitrypsin related emphysema. Am J Respir Cell Mol Biol 2004;31:133–9.

[48] Zhou A, Stein PE, Huntington JA, Sivasothy P, Lomas DA, Carrell RW. How small peptides block and reverse serpin polymerization. J Mol Biol 2004;342:931–41.

[49] Nyon MP, Prentice T, Day J, Kirkpatrick J, Sivalingam GN, Levy G, et al. An integrative approach combining ion mobility mass spectrometry, X-ray crystallography, and nuclear magnetic resonance spectroscopy to study the conformational dynamics of α1 -antitrypsin upon ligand binding. Protein Sci 2015;24:1301–12.

[50] Elliott PR, Pei XY, Dafforn TR, Lomas DA. Topography of a 2.0Å structure of a1-antitrypsin reveals targets for rational drug design to prevent conformational disease. Protein Sci 2000;9:1274–81.

[51] Mallya M, Phillips RL, Saldanha SA, Gooptu B, Brown SC, Termine DJ, et al. Small molecules block the polymerisation of Z a1-antitrypsin and increase the clearance of intracellular aggregates. J Med Chem 2007;50:5357–63.

[52] Berthelier V, Harris JB, Estenson KN, Baudry J. Discovery of an inhibitor of Z-alpha1 antitrypsin polymerization. PLoS One 2015;10:e0126256.

[53] Irving JA, Miranda E, Haq I, Perez J, Kotov VR, Faull SV, et al. An antibody raised against a pathogenic serpin variant induces mutant-like behaviour in the wild-type protein. Biochem J 2015;468:99–108.

[54] Ordóñez A, Pérez J, Tan L, Dickens JA, Motamedi-Shad N, Irving JA, et al. A single-chain variable fragment intrabody prevents intracellular polymerisation of Z a1-antitrypsin. FASEB J 2015;29:2667–78.

[55] Brigham KL, Lane KB, Meyrick B, Stecenko AA, Strack S, Cannon DR, et al. Transfection of nasal mucosa with a normal alpha1-antitrypsin gene in alpha1-antitrypsin-deficient subjects: comparison with protein therapy. Hum Gene Ther 2000;11:1023–32.

[56] Brantly ML, Spencer LT, Humphries M, Conlon TJ, Spencer CT, Poirier A, et al. Phase I trial of intramuscular injection of a recombinant adeno-associated virus serotype 2 alphal-antitrypsin (AAT) vector in AAT-deficient adults. Hum Gene Ther 2006;17:1177–86.

[57] Brantly ML, Chulay JD, Wang L, Mueller C, Humphries M, Spencer LT, et al. Sustained transgene expression despite T lymphocyte responses in a clinical trial of rAAV1-AAT gene therapy. Proc Natl Acad Sci USA 2009;106:16363–8.

[58] Flotte TR, Trapnell BC, Humphries M, Carey B, Calcedo R, Rouhani F, et al. Phase 2 clinical trial of a recombinant adeno-associated viral vector expressing α1-antitrypsin: interim results. Hum Gene Ther 2011;22:1239–47.

[59] Li C, Xiao P, Gray SJ, Weinberg MS, Samulski RJ. Combination therapy utilizing shRNA knockdown and an optimized resistant transgene for rescue of diseases caused by misfolded proteins. Proc Natl Acad Sci USA 2011;108:14258–63.

[60] Cruz PE, Mueller C, Cossette TL, Golant A, Tang Q, Beattie SG, et al. In vivo post-transcriptional gene silencing of alpha-1 antitrypsin by adeno-associated virus vectors expressing siRNA. Lab Invest 2007;87:893–902.

[61] Sehgal A, Vaishnaw A, Fitzgerald K. Liver as a target for oligonucleotide therapeutics. J Hepatol 2013;59:1354–9.

[62] Chan JH, Lim S, Wong WS. Antisense oligonucleotides: from design to therapeutic application. Clin Exp Pharmacol Physiol 2006;33:533–40.

[63] Guo S, Booten SL, Watt A, Alvarado L, Freier SM, Teckman JH, et al. Using antisense technology to develop a novel therapy for α-1 antitrypsin deficient (AATD) liver disease and to model AATD lung disease. Rare Dis 2014;2:e28511.

[64] Guo S, Booten SL, Aghajan M, Hung G, Zhao C, Blomenkamp K, et al. Antisense oligonucleotide treatment ameliorates alpha-1 antitrypsin-related liver disease in mice. J Clin Invest 2014;124:251–61.

[65] Sehgal A, Blomenkamp KS, Qian K, Simon A, Haslett P, Barros S, et al. Preclinical evaluation of ALN-AAT to ameliorate liver disease associated with alpha-1-antitrypsin deficiency. Gastroenterology 2015;148:S-975.

[66] Flotte TR, Mueller C. Gene therapy for alpha-1 antitrypsin deficiency. Hum Mol Genet 2011;20:R87–92.

[67] Mueller C, Tang Q, Gruntman A, Blomenkamp K, Teckman J, Song L, et al. Sustained miRNA-mediated knockdown of mutant AAT with simultaneous augmentation of wild-type AAT has minimal effect on global liver miRNA profiles. Mol Ther 2012;20:590–600.

[68] Brebner JA, Stockley RA. Recent advances in α-1-antitrypsin deficiency-related lung disease. Expert Rev Respir Med 2013;7:213–29.

[69] Takahashi K, Yamanaka S. Induction of pluripotent stem cells from mouse embryonic and adult fibroblast cultures by defined factors. Cell 2006;126:663–76.

[70] Yu J, Vodyanik MA, Smuga-Otto K, Antosiewicz-Bourget J, Frane JL, Tian S, et al. Induced pluripotent stem cell lines derived from human somatic cells. Science 2007;318:1917–20.

[71] Takahashi K, Tanabe K, Ohnuki M, Narita M, Ichisaka T, Tomoda K, et al. Induction of pluripotent stem cells from adult human fibroblasts by defined factors. Cell 2007;131:861–72.

[72] Takahashi K, Yamanaka S. A decade of transcription factor-mediated reprogramming to pluripotency. Nat Rev Mol Cell Biol 2016;17:183–93.

[73] Rashid ST, Corbineau S, Hannan N, Marciniak SJ, Miranda E, Alexander G, et al. Modeling inherited metabolic disorders of the liver using human induced pluripotent stem cells. J Clin Invest 2010;120:3127–36.

[74] Wilson AA, Ying L, Liesa M, Segeritz CP, Mills JA, Shen SS, et al. Emergence of a stage-dependent human liver disease signature with directed differentiation of alpha-1 antitrypsin-deficient iPS cells. Stem Cell Rep 2015;4:873–85.

[75] Yusa K, Rashid ST, Strick-Marchand H, Varela I, Liu PQ, Paschon DE, et al. Targeted gene correction of a1-antitrypsin deficiency in induced pluripotent stem cells. Nature 2011;478:391–4.

[76] Klionsky DJ, Emr. SD. Autophagy as a regulated pathway of cellular degradation. Science 2000;290:1717–21.

[77] Kaushal S, Annamali M, Blomenkamp K, Rudnick D, Halloran D, Brunt EM, et al. Rapamycin reduces intrahepatic alpha-1-antitrypsin mutant Z protein polymers and liver injury in a mouse model. Exp Biol Med 2010;235:700–9.

[78] Marciniak SF, Lomas D. Alpha1-antitrypsin deficiency and autophagy. N Engl J Med 2010;363:1863–4.

[79] Williams RS, Cheng L, Mudge AW, Harwood AJ. A common mechanism of action for three mood-stabilizing drugs. Nature 2002;417:292–5.

[80] Sarkar S, Floto RA, Berger Z, Imarisio S, Cordenier A, Pasco M, et al. Lithium induces autophagy by inhibiting inositol monophosphatase. J Cell Biol 2005;170:1101–11.

[81] Hidvegi T, Ewing M, Hale P, Dippold C, Beckett C, Kemp C, et al. An autophagy-enhancing drug promotes degradation of mutant alpha1-antitrypsin Z and reduces hepatic fibrosis. Science 2010;329:229–32.

[82] Sardiello M, Palmieri M, di Ronza A, Medina DL, Valenza M, Gennarino VA, et al. A gene network regulating lysosomal biogenesis and function. Science 2009;325:473–7.

[83] Pastore N, Blomenkamp K, Annunziata F, Piccolo P, Mithbaokar P, Maria Sepe R, et al. Gene transfer of master autophagy regulator TFEB results in clearance of toxic protein and correction of hepatic disease in alpha-1-anti-trypsin deficiency. EMBO Mol Med 2013;5:397–412.

[84] Pastore N, Ballabio A, Brunetti-Pierri N. Autophagy master regulator TFEB induces clearance of toxic SERPINA1/a-1-antitrypsin polymers. Autophagy 2013;9:1094–6.

[85] Tsvetkov AS1, Miller J, Arrasate M, Wong JS, Pleiss MA, Finkbeiner S. A small-molecule scaffold induces autophagy in primary neurons and protects against toxicity in a Huntington disease model. Proc Natl Acad Sci USA 2010;107:16982–7.

[86] Shoji-Kawata S, Sumpter R, Leveno M, Campbell GR, Zou Z, Kinch L, et al. Identification of a candidate therapeutic autophagy-inducing peptide. Nature 2013;494:201–6.

[87] Blanco I. Use of alpha-1 antitrypsin for the preparation of medicaments for the treatment of fibromyalgia. US7,291,595B2; 2007.

[88] Lewis EC, Dinarello CA, Shapiro L. Methods and compositions for treatment of graft rejection and promotion of graft survival. US20,120,045,460A1; 2011.

[89] Lewis EC, Gottlieb P, Dinarello CA, Shapiro L, Michels A. Compositions, methods and uses for treatment of type 1 diabetes. US20,120,225,813A1; 2012.

[90] Shapiro L. Methods and compositions for treating diabetes. US20,120,040,913A1; 2012.

[91] Quintana AG, Martin JA. Use of alpha-1-antitrypsin for the preparation of drugs for the treatment of chronic fatigue syndrome. US9,205,134B2; 2012.

[92] Lewis E. Combined therapy of alpha-1-antitrypsin and temporal t-cell depletion for preventing graft rejection. US20,150,010,581A1; 2013.

[93] Shapiro L. Compositions of, and methods for, alpha-1 antitrypsin Fc fusion molecules. US20,140,370,036A1; 2014.

[94] Shapiro L. Compositions of, and methods for, alpha-1 antitrypsin Fc fusion molecules. US8,633,305B2; 2014.

[95] Shafiro L, Lewis EC, Dinarello CA. Compositions and methods related to graft versus host disease and treatments thereof. US20,140,127,201A1; 2014.

[96] Crapo JD, Marcondes AMQ, Deeg HJ. Method for ameliorating radiation exposure effects with alpha-1 antitrypsin. US8,980,574B2; 2015.

Index

B

C

D

E

F

G

H

Q

R

S